ECLIPSE

SECOND EDITION

Bryan Brewer

earth view

Seattle, Washington

Eclipse
Second Edition

ISBN 0-932898-91-2

Earth View Inc.
6514 18th Avenue NE
Seattle, WA 98115
USA

Designed by Patrick Jankanish, Seattle, WA

Printed in the United States of America

First printing: February 1991

HOW TO ORDER ADDITIONAL COPIES OF THIS BOOK

For each copy you order, send a check or money order for $17.95 ($14.95 retail price plus $3.00 shipping and handling) to Earth View Inc., at the address shown above.

Sorry, no phone or credit card orders.

to Molly

❧

As in the soft and sweet eclipse,
When soul meets soul on lover's lips.

❧

Shelley, *Prometheus Unbound*

Contents

3. How to Observe an Eclipse

Epilogue: The Future

List of Illustrations

List of Maps

List of Tables

**Based on data provided courtesy of Fred Espenak,
NASA/Goddard Space Flight Center*

Preface to the Second Edition

A total solar eclipse generates a lot of excitement from many quarters. The local inhabitants whose land is destined to be darkened in daytime brace for an onslaught of visitors, while at the same time they try to fathom the meaning of the event for themselves. Astronomers, eclipse chasers, and adventure travelers from around the globe plan years in advance to take up position in the Moon's shadow for a few special moments in time. Educators and media reporters strive to provide accurate information and safe viewing precautions without diminishing the grandeur of the spectacle.

All the excitement about the July 11, 1991, total solar eclipse has led me to issue this new edition of *Eclipse*. Since witnessing totality on February 26, 1979, I can testify that it is an experience to be cherished. As I said in the preface to the first edition: "For a brief moment we can step out of our ordinary world and into the realm of time and space on a truly cosmic scale. It is a chance to see a stunning view of the star at the center of our solar system and to gain some perspective on our place in the universe."

My purpose in writing this second edition is to help maximize your enjoyment and appreciation of that brief moment. Explanations and diagrams will add to your understanding of eclipses and how they happen. Stories of adventure and discovery will bring to life the rich history of these events. Practical tips will help you prepare for the big day. Photographs of eclipse phenomena will give you a sense of the awesome beauty of this experience of a lifetime. And information about upcoming eclipses will fuel your travel plans for the future.

The second edition has been updated with new maps and data for the July 11, 1991, event, as well as with eclipse information about Hawaii and Mexico. The "JULY 91" symbol (shown in the left margin) will help you identify those portions of the book that deal with this event.

I want to express some notes of personal gratitude to a few of the many people who have helped me in the creation of this edition: first to Patrick Jankanish, both for his excellent maps, illustrations, and book design, as well as for the fulfilling co-creative relationship we have enjoyed; to Ken Miller, Fred Espenak, Jay Pasachoff, John Bangert, and Alan Fiala, for providing a wealth of data, feedback, and friendly support for this project; to Patrick Steele and Jane Schulman, each for their wise and caring counsel on this entire effort; and especially to my family for their patience with me and the unfolding process of creativity and growth.

Bryan Brewer
Seattle, Washington
January 1991

Foreword to the First Edition

The starry heavens are as much yours to enjoy as they are an astronomer's, which is a basic message of this informative book by Bryan Brewer. When we look outward with awe and wonder at the marvels awaiting us in an unlimited universe, it is well to recall what he tells us here about the roots of our ancient fascination with the sky. Even if you are someone merely making a wish upon the vision of a bit of cosmic rock burning itself out in Earth's atmosphere, you are in tune with that unbroken past.

Eclipses, those positive markers of our relative movement through the void, make a superb focal point for our outward vision. There is a direct line between Alan Shepherd driving a golf ball across the Moon's surface and the Celtic stonemasons whose stellar observatory still stands on the Salisbury plain. From Stonehenge to that first landing on the Moon, the human statement is plain to read: If it moves, we want to know how and why.

When you stand in the shadow of those bluestone menhirs at Stonehenge to mark the moment of the summer solstice, it is not difficult to feel an affinity with those early astronomers. The archaeological evidence that this carefully arranged series of stone circles was also used to predict eclipses appears to be conclusive. They accomplished much with little beyond their own muscles and intellectual inspiration.

That is likely to be the view distant future generations will take when they consider today's accomplishments. We are always primitives where our own far-off descendants are concerned.

Let the computer-written ephemeris tell you then where and when the next eclipse will occur; that prediction is not different in kind from what the Celtic astronomers told their people.

It is, however, different in quality. It is doubtful that you will perform some esoteric ritual to force the dragon (or snake or worm) to disgorge the Sun.

Mysteries remain, though, and we remake our mythology with increasing frequency, partly because we are still creatures of this planet and caught by a racial compulsion to penetrate beyond the regions that we have already mapped.

I like to think of Bryan Brewer's book as a map of the influences and rhythms contained in eclipses. It is well to remember that more than half the Earth's human population still uses astrology as a guide in the making of decisions. There is a possibility that a kernel of truth remains at the core of this ancient belief. We are Earth creatures. It would be remarkable if the rhythms that influence this planet where we evolved produced no effects on our flesh comparable to the influences upon our religions and philosophies. When we look at the heavens, we look at a cosmic clock that has marked every evolutionary development upon this mundane surface. That clock is still ticking, as the eclipse reminds us.

Frank Herbert
(1920-1986)
Port Townsend, Washington
October 1978

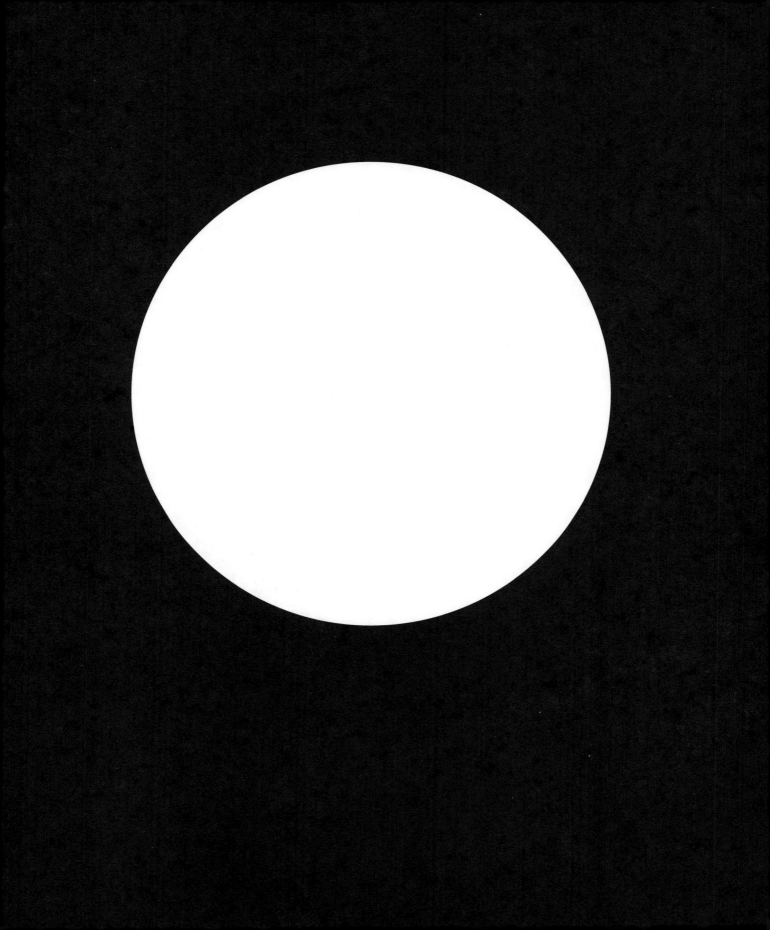

INTRODUCTION

Into the Mouth of the Dragon

It's Thursday morning, July 11, 1991—the day of the eclipse. It is now shortly after sunrise and the gentle morning breezes blow cool in the humid air. As you pace about with anticipation, you begin to wonder if this trip was really worth it. You came quite a distance just to witness a few minutes of darkness. Just as your mind starts to stir up feelings of regret, you hear a shout of excitement.

"First contact! The eclipse has begun."

You rush over to where the eclipse viewer has been set up. People are looking at the image of the Sun projected onto the screen. And sure enough, there on one side of the bright disk, a tiny bit of the Sun has been covered up. You watch for a few minutes and you notice how it's changing: the Moon, slowly moving across the sky, is gradually blocking out more and more of the face of the Sun.

The timing of the eclipse was perfect, beginning within several seconds of the exact moment predicted for this location. Your group selected this site near the Kohala Coast on the Big Island of Hawaii many months ago. You wanted a place near the center of the eclipse path for maximum duration of totality, with an unobstructed view of the morning Sun. After this one, there won't be another total solar eclipse visible from the United States for 26 years. And now, in a little less than an hour, all your plans and expectations will come to a climax in this once-in-a-lifetime event.

The feeling of excitement in the air begins to grow. Every few minutes you check the progress of the Moon inching across the image of the Sun. The Sun appears as a smaller and smaller crescent of light.

© 1990 Roger Ressmeyer/Starlight

Deep sky colors surround the Moon's shadow during total solar eclipse photographed above the clouds at 37,000 feet (July 31, 1981)

For most of the hour after the beginning of the eclipse you barely notice the decrease in sunlight. But as the time approaches for the beginning of totality, the landscape turns quickly darker. A growing sense of uneasiness seems to stir up twinges of an unspoken, primitive fear in all those present to witness this event.

As the narrow crescent of sunlight starts to disappear, little specks of light hang on for a few seconds more. And then, the dark shadow of the Moon rushes over you at incredible speed. The sky is suddenly dark and the corona surrounding the Sun bursts into view.

The delicate light of the solar corona is the crowning glory of the few minutes of totality (March 7, 1970)

Everything is silent. All eyes are held captive by this breathless spectacle in the sky. The irregular shape of the pearly white corona is spread behind the blackened Moon. Your eyes follow the wispy streamers of light extending out into the unreal darkness of the morning sky.

The sunlight from beyond the shadow casts a reddish glow near the horizon. You look around to notice how strangely things appear in the pale illumination of this eerie light. Birds have stopped singing; plants and animals react as if night has fallen. The sudden darkness seems to bring time and Nature to a quiet halt.

A spectacular solar prominence punctuates the inner corona of the Sun (annular eclipse May 30, 1984)

Precious seconds pass as you take in as much as you can of the beauty of these special few minutes in time. It is dark enough to see some stars and possibly a planet or two. Perhaps a few bright red solar prominences rise from the surface of the Sun, these arching flamelike eruptions punctuating the aura of light shining around the dark disk of the Moon. But one thing dominates the sky: the delicate corona—the halo of our Sun—shows its glory for these brief moments, giving you a vision never to be forgotten.

And then, as suddenly as it began, it's over. The shadow passes on and the sunlight returns. The excitement of totality is replaced by a soothing calm. No one talks much at first. As the Moon gradually uncovers more and more of the morning Sun, you are quiet. You want to savor the freshness of the experience, at a loss for words to explain it. Yet somewhere deep inside you have a feeling. An unexpected intimacy with the awesome and relentless forces of Nature has somehow become yours. And you sense that you may never feel quite the same again.

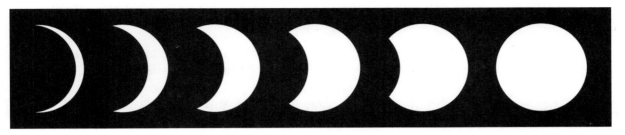

The darkening of the Sun in the middle of the day will always seem an unnatural event. Even today with our scientific understanding of the Earth and space, it still can't help but seem a little frightening to watch the Sun disappear and leave us shrouded in midday darkness. Fortunately, the Sun's "abandoning" of the sky during an eclipse is only temporary. (The word "eclipse" comes from the Greek word meaning "abandonment.") Daylight returns to reassure us, just as it has done for eclipses observed from every corner of the Earth for time immemorial.

The people of many cultures from all parts of the globe have developed various myths or legends about eclipses. Many believe that an eclipse is an omen of some natural disaster or the death or downfall of a ruler. Another pervasive myth involves an invisible dragon or other demon who devours the Sun during an eclipse. Many cultures have developed superstitions about how to counteract the effects of an eclipse. The Chinese would produce great noise and commotion (drumming, banging on pans, shooting arrows into the sky, and the like) to frighten away the dragon and restore daylight. In India, people may immerse themselves in water up to their necks, believing this act of worship will help the Sun and Moon defend themselves against the dragon. In Japan, the ancient custom is to cover wells during an eclipse to prevent poison from dropping into the water from the darkened sky.

This ominous view of eclipses is not the only one. In Tahiti, for example, an eclipse was seen as the lovemaking of the Sun and the Moon. Even to this day, the Eskimos, Aleuts, and Tlingits of Arctic North America believe an eclipse shows divine providence: the Sun and the Moon temporarily leave their places in the sky to check that things are going all right on Earth.

Eclipses Over Hawaii

In Hawaii, where the eclipse of July 11, 1991, first touches land, the ancient native culture embraced the unfavorable view of eclipses. The early Hawaiians believed that attacks by evil gods on the Sun or Moon caused eclipses. In an account of the previous total solar eclipse visible in Hawaii (August 7, 1850), the local newspaper referred to the superstitions of the natives who "…have associated an eclipse, of either the sun or the moon, with the death of their chiefs…" Indeed, this ominous interpretation of unexpected darkness seems quite natural for a people who had no knowledge (as far as we know) of the recurrence of eclipse cycles.

The absence in Hawaiian culture of knowledge of eclipse cycles, as well as the lack of a detailed mythology concerning eclipses, is not difficult to understand. Without a written language, the early Hawaiians could rely only on imprecise oral history to preserve information from one generation to the next. Other ancient cultures that did discover eclipse cycles (such as the Babylonians and the Mayans) could analyze hundreds of years' worth of written astronomical observations. Furthermore, a culture limited to a small isolated geographic region such as Hawaii would not experience very many total solar eclipses. The maps to the left show three consecutive total eclipse tracks over the Hawaiian Islands. (The average frequency is about once every one hundred years.) It makes perfect sense that the early Hawaiians concentrated their astronomical efforts on developing practical and formidable long-range navigational skills rather than on cataloging and analyzing rare conjunctions of the Sun and the Moon.

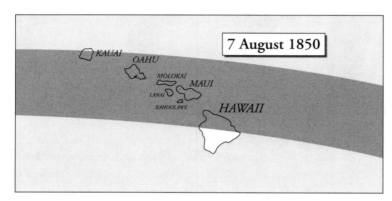

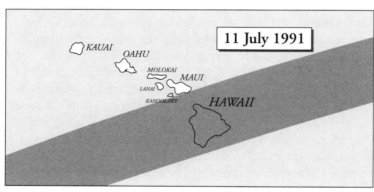

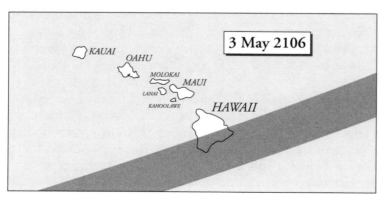

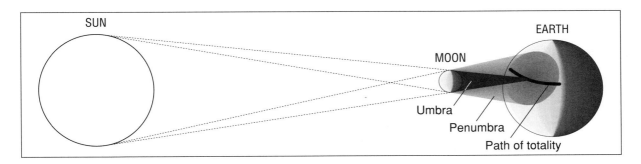

Basic Eclipse Facts

The Moon travels in its orbit around the Earth every 29½ days. An eclipse of the Sun takes place whenever the new Moon, passing right between the Sun and the Earth, casts its shadow on our planet's surface. (This doesn't happen every month; usually the new Moon passes slightly above or below the line between the Sun and the Earth.) From anywhere within the dark cone of the complete shadow, called the umbra (from Latin for "shade"), the Sun will appear in total eclipse. When viewed from the area covered by the larger cone of the penumbra (literally, "almost shade"), the Sun is seen in partial eclipse.

The partial phases of a solar eclipse are normally visible from a broad area of the Earth as wide as 5,000 miles. A total eclipse, on the other hand, can be seen only from the narrow "path of totality" where the Moon's umbra moves across the Earth. This path, sometimes as wide as 200 miles or more, covers only about one-half of one percent of the Earth's surface. Because there are fewer than 70 total eclipses per century, the chance to see one is for most of us a once-in-a-lifetime event.

It is remarkable that total solar eclipses even occur at all. They are possible because the Sun and the Moon appear from the Earth to be about the same size in the sky. The Sun, whose diameter is about 400 times that of the Moon, happens to be about 400 times as far away from the Earth. This condition permits the Moon to just barely cover up our view of the Sun. In fact, if the Moon's diameter (2,160 miles) were just 140 miles less, it would not be large enough to ever completely cover the Sun: a total solar eclipse would never happen anywhere on Earth!

Fortunately, eclipses do occur. Once every year and a half on the average, Nature permits a view of the beautiful corona surrounding the Sun. For ancient people this spectacle of the sky must have had a forceful effect on their consciousness. From the beginnings of written history humankind has recorded eclipses and attempted to understand them. And now, recent investigations have pushed that history further back in time: the intelligent plan of an ancient stone monument, whose construction began as early as 2400 B.C., shows that it could have been used for eclipse predictions.

CHAPTER 1

Eclipses Throughout the Ages

Stonehenge: Eclipse Computer?

Every year on the first day of summer, the Sun rises at a point that is farther north than on any other day of the year. At the ruins of Stonehenge in England, this solstice sunrise appears on the horizon in direct alignment with the upright heel stone. This is the most outstanding feature of this ancient monument, built during the same era as the Great Pyramid of Egypt. There is little doubt that the builders of Stonehenge used it to mark this special day as the beginning of each year. By counting the number of days between these annual alignments, they could determine the length of the year. This could serve as a practical calendar to mark holidays and seasonal festivals and to ensure the timely planting and harvesting of crops.

But to predict eclipses, knowledge of two other cycles is required. One of these—the length of the lunar month—is easily determined. It is simply the number of days between one full Moon and the next. This cycle of 29½ days is marked at Stonehenge by two rings of 29 and 30 holes, which together average 29½. The other cycle, however, is of an altogether different character: it is a cycle of rotation of two invisible points in space. The evidence shows that the builders of Stonehenge probably discovered this cycle and could have used it to predict eclipses.

These two invisible points in space are called the lunar *nodes* (from the Latin for "knot"). They are the points where the Moon's orbit, which is tilted at a slight angle, intersects the plane of the Earth's orbit. It would have taken many decades of watching countless risings and settings of the Moon to figure out the cycle of the lunar nodes. This information—which must have been passed on from generation to generation—is preserved at Stonehenge. All the Moon alignments necessary for determining this cycle are marked by massive stones.

Opposite: View to the northeast where the Sun rises on the first day of summer at Stonehenge

Ruins of Stonehenge today

Who were these people who observed this subtle cycle even before the first metal tools were used by humankind? Some have suggested that Stonehenge was built by Druids, but we don't really know much about the builders. We do know that the actual motions of the Sun and the Moon are reflected in the structure of this monument, and we can reason how it may have been used to keep track of these cycles. The number of stones or holes in the ground in the various rings around Stonehenge each represents a certain number of days or years in the cycles. By moving markers (such as stones) around a ring in time with the cycles, the positions of the Sun and Moon—and the two invisible points—can be tracked. (The details of this method are explained in Chapter 2.)

An eclipse can occur only when the Sun is close to being aligned with a node. By using Stonehenge to keep track of the position of the Sun and the nodes, these "danger periods" for eclipses can be predicted. A new (or full) Moon appearing during one of these periods would call for a special vigil to see if the solar (or lunar) eclipse would be visible from Stonehenge. (A lunar eclipse occurs when the full Moon is engulfed in the Earth's shadow; see p. 45.) A total solar eclipse would be a rarity. But the law of averages confirms that either a partial solar eclipse or a lunar eclipse can be seen (weather permitting) from the same point on the Earth about once every year.

Dragons and Serpents: Eclipse Symbols?

Why would eclipses have been so important to the ancient people of Stonehenge? Perhaps they considered the darkening of the Sun or the Moon a fearsome event—a celestial omen of doom or disaster. Many cultures have interpreted eclipses this way. But the sophistication of the astronomy of Stonehenge suggests that the builders had something different in mind. Their understanding of the solar and lunar cycles must have led to a high regard for the cosmic order. Eclipses may have been seen as affirmations of the regularity of these cycles. Or perhaps the unseen lunar nodes formed an element of their religion as invisible gods capable of eclipsing the brightest objects in the heavens.

The idea that Stonehenge may have been a center for some kind of worship has occurred to many. It is not hard to imagine Stone Age people gathering at a "sacred place" at "sacred times" (such as solstices, equinoxes, and eclipses) to reaffirm their religious beliefs through ritual practices. British antiquarian Dr. William Stukeley, who in 1740 was the first to note the summer solstice alignment at Stonehenge, advanced the notion that the monument was built by Druids to worship the serpent. He claimed that Stonehenge and similar stone circles had been serpent temples, which he called "Dracontia." Could this serpent symbolism be related to eclipses? Recall that the key to eclipses is the position of the lunar nodes. The length of time for the Moon to return to a node (about 27.2 days) astronomers call the *draconic month*. (Draco is the Latin word for "serpent" or "dragon.") Perhaps the mythical serpents of Stonehenge and the legendary dragon that eats the Sun are symbols of the same thing: the invisible presence in time and space that eclipses the Sun and the Moon.

Fanciful illustration of early Stonehenge celebration

The Birth of Astronomy

Whatever the reasons for Stonehenge, they are lost in time; the builders left no written records. During this same period in history (between 3000 B.C. and 2000 B.C.), the study of the heavens was developing as a written science in the Middle East. Astronomers in Babylonia and Assyria kept track of time by carefully observing the motions of the Sun and the Moon. They increased the accuracy of their measurements by recording the details of solar and lunar eclipses. As they studied this record of centuries of eclipses, a pattern of repetition began to emerge: eclipses tend to repeat themselves every 18 years, although they recur at different places on the globe. This eclipse cycle, called the *saros*, is used even to this day to make predictions. For example, the July 11, 1991, eclipse is included in a *saros* series. On June 30, 1973, exactly 18 years and 11⅓ days earlier, a solar eclipse took place. Another solar eclipse will occur on July 22, 2009, again after 18 years and 11⅓ days. (See pp. 52–53 and 68–70 for more details on the *saros*.)

The Babylonian discovery of the *saros*, important for eclipse predictions, is not the most famous of their contributions to astronomy. As early as 3000 B.C., they originated the division of the sky into the 12 signs of the zodiac, and the names they gave to each sign are still used. Today these names and symbols (the symbols are of unknown origin) are more familiar to the practice of astrology; but in ancient Babylonia, astronomy and astrology were inseparably connected.

Zodiac sign	Symbol	Ancient "ruling planet"	Zodiac sign	Symbol	Ancient "ruling planet"
Aries	♈	Mars	Libra	♎	Venus
Taurus	♉	Venus	Scorpio	♏	Mars
Gemini	♊	Mercury	Sagittarius	♐	Jupiter
Cancer	♋	Moon	Capricorn	♑	Saturn
Leo	♌	Sun	Aquarius	♒	Saturn
Virgo	♍	Mercury	Pisces	♓	Jupiter

The religion of the Babylonians was based on the belief that earthly affairs were influenced by the motions of heavenly bodies. It was the duty of the astrologer-priests to keep watch on the skies and warn of any disasters that might be signaled. They developed an elaborate system of celestial omens to "divine" the future. It seems that the Babylonians were more interested in discerning meaning from the paths of the Sun, the Moon, and the planets across the background of the stars than in discovering the secrets of the physical world around them. They also believed that each sign of the zodiac was influenced by one of the "ruling planets," which imparted its qualities to events related to that sign.

The Babylonians also contributed to the establishment of the seven-day week. As the 12 signs of the zodiac related to the 12 lunar months in a year, the seven days in a week probably come from the quarter phases of the lunar month. Also, ancient astronomers/astrologers recognized seven "planets" (including the Sun and the Moon); they associated each planet, personified as a celestial deity, with a day of the week. The present names for the days of the week are derived from this same scheme, using Roman or Norse names for the planets and deities.

Celestial deities from which the days of the week originated (celestial bodies are keyed by number to the illustration below)

Norse deity	English name for the day of the week	Celestial body	Spanish name for the day of the week
	Sunday	1. Sun	Domingo
	Monday	2. Moon	Lunes
Tiw (god of war)	Tuesday	3. Mars	Martes
Woden (chief god)	Wednesday	4. Mercury	Miércoles
Thor (god of thunder)	Thursday	5. Jupiter	Jueves
Freya (goddess of marriage)	Friday	6. Venus	Viernes
	Saturday	7. Saturn	Sabado

The Winged Sun over Egypt

As the Babylonians were developing the science of astronomy, the ancient Egyptian civilization was flourishing. Pyramids, temples, and tombs attest to the high state of development of their art and technology. They measured the length of the year by observing the rising of Sirius, the brightest star in the sky. The Great Pyramid at Giza is aligned to the four points of the compass; it was built with a passageway in alignment with the star that was then the pole star, Alpha Draconis. There is no doubt that the Egyptians watched the heavens. The clear skies of the Nile Valley were ideally suited for celestial observation. Yet no one has found a single reference to an eclipse, either of the Sun or the Moon, in all of ancient Egyptian history.

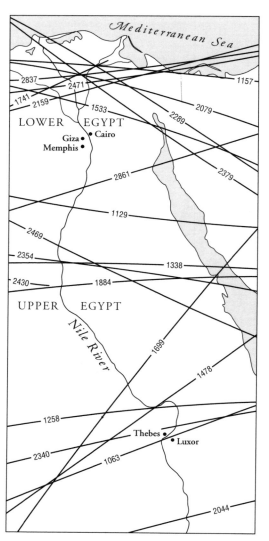

This apparent gap in Egyptian astronomy has puzzled many historians. Was Egypt shortchanged on total solar eclipses? Far from it. The accompanying map shows the central lines of all total eclipses across the Nile Valley in the second and third millennia B.C. The solar corona was visible from somewhere in ancient Egypt during this period on an average of once every 75 years. It is hard to imagine that the spectacular recurrence of total solar eclipses could go unrecorded, especially by a culture that so worshipped the Sun.

Perhaps the view of totality was preserved in symbolic form. The solar corona has a distinctive appearance during some eclipses. The size and shape of this halo around the Sun varies over a cycle of 11 years.

Total Solar Eclipses in the Nile Valley 3000 B.C.–1000 B.C.

23 Mar. **2861** B.C.	20 Apr. **2044** B.C.
19 Nov. **2837** B.C.	15 Sep. **1884** B.C.
1 Apr. **2471** B.C.	21 Dec. **1741** B.C.
2 Sep. **2469** B.C.	16 Apr. **1699** B.C.
25 Jul. **2430** B.C.	9 May **1533** B.C.
27 Oct. **2379** B.C.	1 Jun. **1478** B.C.
25 Jun. **2354** B.C.	14 May **1338** B.C.
23 Mar. **2340** B.C.	27 Jul. **1258** B.C.
20 Dec. **2289** B.C.	19 Aug. **1157** B.C.
29 Jun. **2159** B.C.	14 Feb. **1129** B.C.
11 Sep. **2079** B.C.	31 Jul. **1063** B.C.

Central lines and years (B.C.) of total solar eclipse paths across ancient Egypt

Equatorial streamers seen at July 29, 1878, eclipse at Pikes Peak

(This is the sunspot cycle explained later in this chapter.) During the minimum phase of this cycle, the brightness of the corona is less intense, but extending to either side are long streamers of light. Because these equatorial streamers are so faint, they are difficult to photograph. Yet in clear skies they are plainly visible to the naked eye during an eclipse.

It is not difficult to see the similarity between these eclipse streamers and the symbolic wings of the Egyptian Sun. The drawing above was made by Samuel P. Langley from the summit of Pikes Peak during the eclipse of July 29, 1878. The long equatorial streamers are well defined. The symbol below is the winged disk of the Sun; it was one of the earliest solar representations in Egypt. It appears above the entrances of many tombs and temples and is said to commemorate the victory of light over darkness. Sometimes the symbol includes the heads of two serpents and the horns of a goat, also solar symbols.

Could this view of the eclipsed Sun be the ancient source of this widespread symbol? English astronomer E. W. Maunder put it this way:

> …there can be little doubt that the Sun was regarded partly as a symbol, partly as a manifestation of the unseen, unapproachable Divinity. Its light and heat, its power of calling into active exercise the mysterious forces of germination and ripening, and the universality of its influence, all seemed the fit expressions of the yet greater powers which belonged to the Invisible.

> What happened in a total solar eclipse? For a short time that which seemed so perfect a divine symbol was completely hidden. The light and heat, the two great forms of solar energy, were withdrawn, but something took their place. A mysterious light of mysterious form, unlike any other light, unlike any other single form, was seen in its place. Could they fail to see in this a closer, a more intimate revelation, a more exalted symbolism of the Divine Nature and Presence?

Knowledge, vol. XX, p. 9, January 1897

Winged solar disk adorns the top of King Tutankhamun's ceremonial chair

Eclipses in History and Literature

Stonehenge, Babylonia, Egypt—each culture developed a unique approach to eclipses. But only the Babylonians discovered the long-range cycle, the *saros*. An eclipse cycle can also be used to go backward in time. This technique has proven useful to historians in fixing exact dates of past events.

Numerous systems were used in ancient civilizations to keep track of the passage of time. Typically, routine happenings would be recorded as so many days, months, or years after some memorable event such as the crowning of a ruler, a natural catastrophe, or other momentous occasion. Often there would be no indication of exactly when the reference event took place. If an eclipse was described in the record of events, it could be compared with actual eclipses that were known to have happened near the time and place in question. If there were only one eclipse that fit the description, then the dates could be fixed with certainty. Many historical chronologies have been verified or compared using this method.

The earliest record of a solar eclipse comes from ancient China. The date of this eclipse, usually given as October 22, 2134 B.C., is not certain. Historians know the account was written sometime within a period of about two hundred years. During that time there were several total eclipses visible in China. The 2134 B.C. eclipse is simply the best guess.

The date of an eclipse referred to in the Bible is known for certain: " 'And on that day,' says the Lord God, 'I will make the Sun go down at noon, and darken the Earth in broad daylight'." (*Amos* 8:9) "That day" was June 15, 763 B.C. The date of this eclipse is confirmed by an Assyrian historical record known as the *Eponym Canon*. In Assyria, each year was named after a different ruling official and the year's events were recorded under that name in the *Canon*. Under the year corresponding to 763 B.C., a scribe at Nineveh recorded this eclipse and emphasized the importance of the event by drawing a line across the tablet. These ancient records have allowed historians to use eclipse data to improve the chronology of early Biblical times.

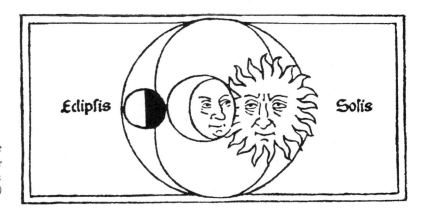

Solar eclipse woodcut from Johannes de Sacrobusco's Opus Sphæricum *(1482)*

Wars and Earthquakes

What is probably the most famous eclipse of ancient times ended a five-year war between the Lydians and the Medes. These two Middle Eastern armies were locked in battle when "the day was turned into night." The sight of this total solar eclipse (the date is fixed as May 28, 585 B.C.) was startling enough to cause both nations to stop fighting at once. They agreed to a peace treaty and cemented the bond with a double marriage. The eclipse was predicted by Thales, the celebrated Greek astronomer and philosopher, but the prediction was probably not known to the warring nations.

Battle between Lydians and Medes (585 B.C.) halted by total solar eclipse

The lunar eclipse of August 27, 413 B.C., had a different effect on the outcome of a battle in the Peloponnesian War. The Athenians were ready to move their forces from Syracuse when the Moon was eclipsed. The soldiers and sailors were frightened by this celestial omen and were reluctant to leave. Their commander, Nicias, consulted the soothsayers and postponed the departure for twenty-seven days. This delay gave an advantage to their enemies, the Syracusans, who then defeated the entire Athenian fleet and army, and killed Nicias.

The spectacle of an eclipse, which had a powerful effect on decisions in battle, was equally impressive to ancient poets. A fragment of a lost poem by Archilochus contains the words:

> Nothing there is beyond hope, nothing that can be sworn impossible, nothing wonderful, since Zeus, father of the Olympians, made night from mid-day, hiding the light of the shining Sun, and sore fear came upon men.

This has been identified as a description of the total solar eclipse of April 6, 648 B.C. Another eclipse reference (from the Bible) goes like this:

> And I behold when he had opened the sixth seal, and lo, there was a great earthquake; and the Sun became black as sackcloth of hair, and the Moon became as blood.
> —*Revelation* 6:12

This compelling passage is only one of a number of literary and historical connections between eclipses and earthquakes. The Greek historian Thucydides, in writing about the Peloponnesian War, remarked about "earthquakes and eclipses of the Sun which came to pass more frequently than had been remembered in former times." On another occasion he noted "…there was an eclipse of the Sun at the time of a new Moon, and in the early part of the same month an earthquake." Another Greek writer, Phlegon, reported the following events:

> In the fourth year of the 202nd Olympiad, there was an eclipse of the Sun which was greater than any known before and in the sixth hour of the day it became night; so that stars appeared in the heaven; and a great earthquake that broke out in Bithynia destroyed the greatest part of Nicæa.

This interest in linking the two types of events by coincidence may have been attempts to derive some order out of the unpredictability of earthquakes, possibly a carryover from the celestial omens of the Babylonians. Although no scientific connection between earthquakes and eclipses has been proven, the frequency of occurrence of both types of events will continue to produce intriguing coincidences. The earthquake in Iran on September 16, 1978, the most devastating one of that year and which killed more that 25,000 people, occurred just 3½ hours before a total lunar eclipse was visible there.

Mystified by the Moon

These kinds of ominous events have played important parts in human history. When English poet John Milton, in *Paradise Lost*, wrote these lines

> As when the Sun, new risen,
> Looks through the horizontal misty air,
> Shorn of his beams, or from behind the Moon,
> In dim eclipse, disastrous twilight sheds
> On half the nations and with fear of change
> Perplexes monarchs

he may have been thinking of Charlemagne's son, Emperor Louis. This European ruler was so "perplexed" by the five minutes of totality he witnessed during the eclipse of May 5, 840, that he died (some say of fright) shortly thereafter. The fighting for his throne ended three years later with the historic Treaty of Verdun, which divided Europe into the three major areas we know today as France, Germany, and Italy.

Solar eclipses perplexed the common people as well. Medieval historian Roger of Wendover reported on the total eclipse of May 14, 1230, which occurred early in the morning in Western Europe: "…and it became so dark that the labourers, who had commenced their morning's work, were obliged to leave it, and returned again to their beds to sleep; but in about an hour's time, to the astonishment of many, the Sun regained its usual brightness." This was during the Dark Ages and an understanding of eclipses was not common knowledge.

Mark Twain used this ignorance of eclipses as an element of the plot in *A Connecticut Yankee in King Arthur's Court*. The hero of the novel, Hank Morgan, is mysteriously transported backward in time to Medieval England. He finds himself about to be burned at the stake on a day when he knows a solar eclipse will occur. He "foretells" the event, claiming to have magical powers over the Sun. "The rim of black spread slowly into the Sun's disk, …the multitude groaned with horror to feel the cold uncanny night breezes and see the stars come out…" Morgan promises to restore the daylight in exchange for his freedom. King Arthur agrees and, of course, the Sun returns. Twain gives the date of the eclipse as June 21, 528. This, however, is literary fiction; no such eclipse took place on or near that date.

A similar sort of deception was actually used by Christopher Columbus during his fourth voyage to the Americas. In 1503, he found himself stranded on the island of Jamaica, his ships damaged beyond repair and his provisions running low. At first he and his crew were able to get food from the natives in trade for baubles and trinkets. But as months passed without rescue, the Jamaicans finally refused to supply any more food. Faced with the prospect of starvation, the great Spanish admiral conceived an ingenious plan.

Columbus knew from his navigational tables that a total eclipse of the Moon would occur on February 29, 1504. He arranged a meeting with the natives that evening to coincide with the beginning of the eclipse. He announced that because God didn't like the way the natives were treating him and his crew, the Almighty had decided to remove the Moon as a sign of his displeasure! Columbus timed his theatrics precisely; no sooner had he proclaimed the Moon's disappearance than the Earth's shadow began to steal across the face of the full Moon.

The natives were terrified. As the light of the Moon faded they pleaded with Columbus to restore it; they would give him all the food he wanted if he would bring back the Moon. Columbus told them he would have to retire to confer with God, which in this case was an hourglass timing the eclipse. Just before the end of the total phase he announced that God had pardoned them and would allow the Moon to return to its place in the sky. And as Columbus knew it would, the Moon reappeared. The grateful natives resumed the supply of food, and Columbus and his crew were eventually rescued and returned to Europe.

Terrified Jamaicans plead with Columbus to restore eclipsed Moon

The Science of Prediction

Today's astronomers are able to predict the precise time and location of solar eclipses. With advance notice of the event and a higher level of scientific understanding among people, there is no need for anyone to be frightened by what should be a marvelous experience of the beauty of Nature. In fact, many people make plans to travel to locations in the path of the Moon's shadow just to witness the spectacle of a total solar eclipse. Astronomers at the U. S. Naval Observatory in Washington, D.C., process the data and publish predictions a year or two in advance for all eclipses. The times and locations (including maps) appear in the annual publication *The Astronomical Almanac*. According to the Observatory astronomer who maintains the computer programs for eclipses, the prediction of the path of totality is accurate to within one or two miles and the timing of the eclipse to within a few seconds.

Eclipse predictions have not always been that accurate. Stonehenge could be used to forecast the day of an eclipse, but not the specific time or place. The ancient Chinese, Babylonians, and Greeks made improvements, but it was not until the 17th century A.D., after Copernicus had shown that the Sun is at the center of the solar system and Newton had formulated the laws of gravity, that eclipse predictions achieved modern accuracy. The actual motion of the Moon is fraught with numerous small disturbances and discrepancies. In 1693 British astronomer Edmond Halley (of comet fame) was the first to notice a small but steady change in the Moon's motion called *secular acceleration*. This simply means that the Moon is slowly gaining speed in its orbit. Modern astronomers use ancient eclipse records, some several thousand years old, to determine the value of this change.

In 1824 a great practical stride was made in eclipse predictions: Prussian astronomer Friedrich Bessel introduced a group of mathematical formulas that greatly simplified the calculation of the positions of the Sun, Moon, and Earth. These "Bessel functions," which are used even today, laid the foundation for a monumental book on eclipses. An Austrian astronomer named Theodor von Oppolzer organized the calculation of all eclipses from 1207 B.C. to 2161 A.D. One year after his death in 1886, his *Canon of Eclipses* was published. It contains the details of the time and place for the 13,200 solar and lunar eclipses for those 34 centuries. Another Austrian astronomer, Friedrich Ginzel, used these data for historical research on eclipses. In 1899 he published his *Special Canon of Solar and Lunar Eclipses*, which shows the references in classical literature to all eclipses between 900 B.C. and 600 A.D. Both of these works have been valuable tools for historians who use eclipses to verify dates in history.

*Map produced by
Edmond Halley
showing the path of
totality across England
on April 22, 1715*

Modern Eclipse Expeditions

Astronomers today travel to all parts of the globe to gather eclipse data, meeting the Moon's shadow wherever it happens to touch the Earth. This has not always been the case. It is only in the last 150 years or so that eclipse expeditions have been in vogue. A notable exception occurred for the total solar eclipse of October 27, 1780: Samuel Williams, professor at Harvard, led an eclipse expedition to Penobscot Bay, Maine. The exceptional part of the story is that this happened during the Revolutionary War and Penobscot Bay lay behind enemy lines. Fortunately, the British granted the party safe passage, citing the interest of science above political differences.

Until the middle of the last century, most of the scientific interest in eclipses concerned the precision of orbital motion. Astronomers used the data from eclipse observations to refine their knowledge of celestial mechanics, which in turn led to more accurate eclipse predictions. Little attention was paid to describing the visible phenomena of total solar eclipses.

This situation changed when the Moon's umbra crossed populated parts of Southern Europe on July 8, 1842. Those who observed totality on that day were rewarded with a magnificent view of the corona and prominences. Francis Baily, an English amateur astronomer, was the first to use the word "corona" as an astronomical term in describing this eclipse. Scientists were stirred to discover more about this halo of light and the "red flames" that appeared around the Moon. Luckily, the technology of photography was beginning to develop at that time. The first successful photograph of the corona was taken at the total eclipse in Northern Europe in 1851. Scientists in 1860 used eclipse photos taken in Spain to show that the solar prominences were definitely part of the Sun and not part of the Moon as some had believed.

Corona seen in 1842

Opposite: 17th century astronomer Hevelius observing a solar eclipse by projection into a darkened room (from his Machina Coelestis, *1673)*

About this time another breakthrough happened that opened entirely new avenues of investigation in solar physics. For many years scientists had noticed a number of thin dark lines in the rainbow spectrum of light from the Sun. In 1859, German physicist Gustav Kirchhoff accounted for their origin: the lines occurred because of the chemical elements present in the Sun. Since each element has its own distinctive set of lines, the chemical composition of the Sun could be derived from its spectrum.

This new technique, called spectroscopy, was first applied to the eclipsed Sun on August 18, 1868. By this time, eclipse expeditions to remote areas of the globe were routine, and many traveled to India and Malaya to see this eclipse. British astronomer Norman Lockyer trained his spectroscope on the solar prominences and discovered a spectral line of a new chemical element. He named it helium (from Greek *helios*, the Sun); this familiar gas was not identified on the Earth until 1895. At the same eclipse, he and French astronomer Pierre Jules Janssen, each working independently, figured out a spectroscopic method for observing the prominences at times other than during an eclipse.

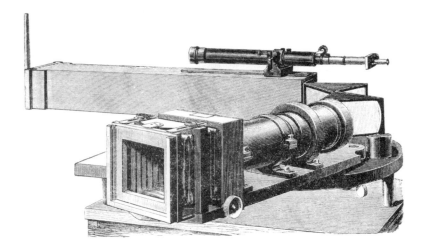

19th century spectroscope used for studying Sun during eclipses

In the following ten years steady advances were made in the spectroscopy and photography of eclipses. Expeditions to America, the Mediterranean, India, South Africa, and Siam yielded new information about the composition of the Sun and the structure of the corona. These expeditions to remote areas presented many challenges to astronomers. Transporting large, sensitive telescopes and other instruments compounded the hardships of global travel a century ago. And once they were set up at a site within the path of totality, cloudy skies during the eclipse could defeat the purpose of the journey.

The expedition to India for the eclipse of December 12, 1871. These British astronomers (Norman Lockyer is seated at the left under the umbrella) made their observations from atop a tower at the old fort at Bekul. A crowd of astonished natives gathered around the tower; in their alarm at the sight of the disappearing Sun, they kindled a fire in preparation for a sacrifice. The astronomers, fearful the smoke would obscure their view, had the police stop the attempted fire-lighting.

ↄↄ

High on her speculative tower
Stood Science waiting for the hour
When Sol was destined to endure
That darkening of his radiant face
Which Superstition strove to chase,
Erewhile, with rites impure.

ↄↄ

Wordsworth, *The Eclipse of the Sun, 1820*

Vol. XXII.—No. 1130.] NEW YORK, SATURDAY, AUGUST 24, 1878. [WITH A SUPPLEMENT PRICE TEN CENTS.

Entered according to Act of Congress, in the Year 1878, by Harper & Brothers, in the Office of the Librarian of Congress, at Washington.

The Sunspot Connection

None of these inconveniences deterred the scientists who traveled to Colorado and Wyoming for the total eclipse of July 29, 1878. The transcontinental railroad had been completed nine years earlier and astronomers were offered half-price fares for the trip from the East Coast. For a short period that summer, obscure towns in the West became centers of scientific activity. Famous astronomers from Europe and all over America turned out to see the eclipse in the clear skies of the Rocky Mountains. Even Thomas Edison (no astronomer himself) was there to test a new invention he claimed could measure the heat of the corona.

A group headed by Samuel P. Langley, later director of the Smithsonian Institution, climbed to the summit of Pikes Peak in Colorado to witness totality on that summer afternoon. The day before the eclipse was ominous; they experienced hail, rain, sleet, snow, and fog! But eclipse day was clear and their perseverance was rewarded by a startling sight: two coronal streamers extending in opposite directions as far as twelve diameters of the Sun. This was much wider than had ever been seen before by scientists.

Although the 1878 corona was very wide, it was actually not as bright as those seen in 1870 and 1871. Astronomers began to suspect that the corona's shape and intensity were related to levels of activity on the Sun. One of the measures of solar activity is the occurrence of sunspots. These dark blotches appear on the Sun's surface, sometimes lasting for many weeks. Some years earlier Heinrich Schwabe, a German amateur astronomer, noted that the average number of sunspots per day varied in a regular cycle of approximately 11 years. Could this cycle be linked to the changing shape of the corona?

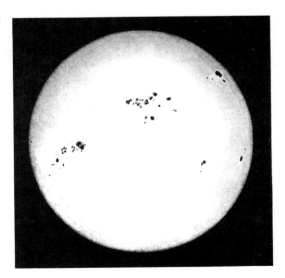

Photograph showing high level of sunspot activity in 1870

Opposite: *The "Great Solar Eclipse" of July 29, 1878*

Observations of the corona over the following decades proved this theory correct. An eclipse that occurs near a low point in the sunspot cycle (as in 1878) reveals a corona that is dimmer than normal but that shows a more detailed structure. The coronal "halo" seen surrounding the black disk of the Moon is somewhat compressed. The long *equatorial streamers* may be seen stretching out from either side of the Sun. Finely detailed *polar plumes* of light curve above and below the dark disk in the sky. Near sunspot maximum (as in 1991), the appearance is just the opposite. The plumes and streamers are less pronounced, but the coronal glow around the Sun is brighter and more expanded. Eclipses occurring at intermediate stages of the cycle exhibit some combination of features of both types.

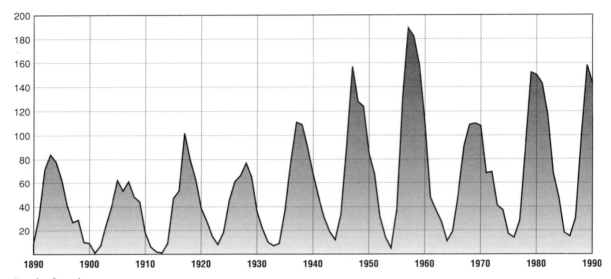

Graph of yearly sunspot numbers

The 1878 eclipse also marked the height of the search for the elusive planet "Vulcan." This was the name given to a small object reportedly seen near the Sun on several occasions; some scientists thought it was a planet that up to then had escaped detection. The calculation of small irregularities in the orbit of Mercury supported this theory. Just 30 years earlier Neptune had been discovered in a similar manner. Astronomers were hoping that the clear skies and the blacked-out Sun would reveal the planet in their telescopes during the eclipse. But it didn't happen. One astronomer did announce he had discovered Vulcan, but he was later proved wrong. The discrepancies in Mercury's orbit are fully accounted for by Einstein's theory of relativity.

Eclipses and Einstein's Theory of Relativity

Because the Sun's light is shielded during an eclipse, some of the brighter stars and planets can be seen in the darkened sky. This fact has enabled astronomers to test part of the theory of relativity. According to Einstein, who proposed the theory in 1915, rays of light should be deflected by a gravitational field. In particular, starlight passing near the Sun should be bent slightly toward the Sun. The only time when stars near the Sun are visible is during a total solar eclipse.

Scientists put the theory to test during eclipse expeditions to Brazil in 1919 and to Australia in 1922. Photographs of stars near the Sun during these eclipses were compared to photographs taken of the same stars several months later when the Sun, in another part of the sky, would have no effect. The difference in position of the stars showed that Einstein was correct. Eclipses became the first tool to crack the door of experimental proof on one of the most profound scientific ideas about the universe.

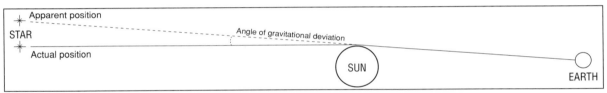

Bending of starlight as observed during an eclipse

Scientists at these "relativity eclipses" were blessed with clear skies. The excitement generated by this new theory led to great plans for observation of totality on September 10, 1923. The path would graze Southern California at a favorable time of year. Weathermen predicted a 90% chance of clear skies. But as fate would have it, the day was overcast and the clouds spoiled all the planned observations.

A year and a half later there was little hope of good weather for the total eclipse in the northeast United States. Yet many places in the path of totality in New York and Connecticut experienced clear skies on January 24, 1925. Millions of people witnessed the eclipse. The southern edge of the path crossed right through New York City. This situation provided a unique opportunity to determine the precise location of the edge of totality during an eclipse.

Astronomers knew beforehand that the edge of the Moon's shadow would cut Riverside Drive in New York City somewhere between 83rd and 110th Streets. To be on the safe side, observers were positioned at every intersection between 72nd and 135th Streets. They were instructed to report whether they had seen the corona (total phase) or only a crescent of the Sun (partial phase). The results were definite: the edge of the umbra passed between 95th and 97th Streets, yielding an accuracy of several hundred feet for a shadow cast a distance of over 200,000 miles.

Chasing the Moon's Shadow

In recent years, great technological advances have occurred in many types of instruments used to gather scientific data during eclipses. Much of this increased sensitivity is lost, however, because of distortion by the lower atmosphere. More accurate data can be gathered at higher elevations, but eclipse paths don't often pass over convenient mountaintops. (A notable exception is the eclipse of July 11, 1991, which passes directly over the cluster of observatories on the summit of Mauna Kea, Hawaii.)

The modern astronomer's solution has been to take to the air. Three distinct advantages result from the use of "flying observatories" in the past few decades. First, flying above the clouds ensures that bad weather will not spoil the occasion. Second, the clarity of the atmosphere at high altitudes provides better results. And third, because an aircraft can fly in the direction the shadow is moving, the effective duration of totality can be lengthened. On June 30, 1973, scientists aboard the supersonic aircraft *Concorde 001* flew in the Moon's shadow across Africa for 74 minutes—ten times longer than an eclipse can ever be observed from the ground.

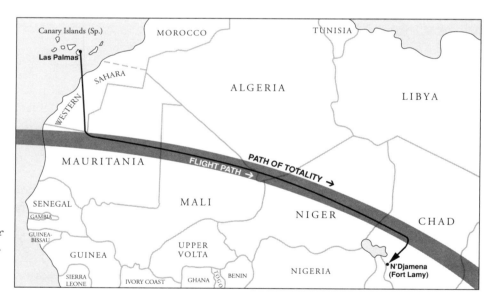

Flight path of Concorde 001 *in the Moon's shadow on June 30, 1973*

But the story doesn't end there. The study of the Sun is also reaching into space. Observations of the solar corona from orbiting spacecraft have expanded our understanding of the Sun into areas previously unexplored. This increase in solar knowledge coincides with a growing public awareness of the Sun as the primal source of energy for our planet. A total solar eclipse provides a magnificent opportunity to personally appreciate the source of life-giving energy at the center of our solar system.

CHAPTER 2

Understanding Eclipses

The Approach of Darkness

A total eclipse begins almost unnoticeably. First contact occurs when the Moon starts its passage across the face of the Sun. At first, only a small "bite" appears on the western edge of the Sun. Gradually, as more and more of the Sun disappears, an interesting effect can be seen: the tiny spots of light shining through the leaves of a tree, for example, show up on the ground as crescent images of the slowly vanishing Sun.

This partial phase of the eclipse leads to totality in about an hour. For most of that time, there is little hint of the approaching darkness. But as the bright area of the Sun is reduced more and more, the increasing darkness becomes noticeable. Daylight fades very quickly in the last few minutes before totality.

Crescent images of partially eclipsed Sun (1900 engraving)

Shadow bands visible just prior to totality (1900 engraving)

While a small crescent of the sun remains in the sky, a curious eclipse phenomenon is often observed. Thin wavy lines of alternating light and dark can be seen moving and undulating in parallel on plain light-colored surfaces. These so-called *shadow bands* are the result of sunlight being distorted by irregularities in the Earth's atmosphere. An open floor or wall is a good place to look for them. A similar effect is seen when the Sun shines through ripples on the surface of the water in a swimming pool; the wavy lines moving on the bottom of the pool resemble the shadow bands of an eclipse.

As the narrow crescent of the Sun finally begins to disappear, tiny specks of light remain visible for a few seconds more. These points of light are spaced irregularly around the disappearing edge of the Sun, forming the appearance of a string of beads around the dark disk of the Moon. These lights are known as *Baily's beads*, named after Francis Baily, the 18th century English amateur astronomer who was the first to draw attention to them.

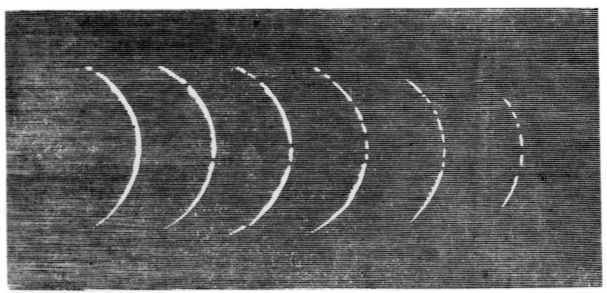

Baily's beads (engraving from the eclipse of July 18, 1860)

Baily's beads would not be possible if the Moon's surface were perfectly smooth. The edge of the Sun is first hidden by the peaks of lunar mountains. The beads are the last few rays of sunlight shining through valleys on the edge of the Moon. Baily's beads make their brief appearance up to 15 seconds before totality. When a single point of sunlight remains, a beautiful "diamond ring" effect is created against the outline of the Moon. This final sparkling instant signals the arrival of the Moon's shadow. The last ray of sunlight vanishes and totality begins.

❧

O dark, dark, dark, amid the blaze of noon,
Irrevocably dark, total eclipse
Without all hope of day.

❧

Milton, *Paradise Regained*

The Spectacle of Totality

Suddenly the sky above is dark. The Moon's shadow, racing along the Earth at speeds up to several thousand miles per hour, brings a swift and dramatic nighttime effect. The sky near the horizon, where the eclipse is not total, still appears bright. This distant scattered light produces a slight reddish glow and unusual shadow effects. This daytime darkness is not quite as black as at night. But its startling onset and unearthly appearance combine to create a unique visual ambience.

In the center of this darkened sky hangs the featured spectacle of the eclipse—the corona of the Sun. This pearly white crown of light shines in all directions around the darkened solar disk. A million times fainter than the Sun itself, the full glory of the corona is visible only during a total solar eclipse.

The corona consists of the ionized gases that form the outer atmosphere of the Sun. Although these gases extend many millions of miles into space, only the corona near the Sun is visible to the naked eye. Wispy plumes and streamers of coronal light reach out distances up to several diameters of the Sun before they fade into darkness.

The corona comes into full view when the leading edge of the Moon blots out the last ray of sunlight, and it remains visible throughout totality. For a few seconds both after the beginning and before the end of totality, a pinkish glow appears at the edge of the Moon. This is light from the Sun's lower atmosphere, the *chromosphere.* Its rosy color ("chromo" means color) comes from its main element, hydrogen.

Extending outward from the chromosphere are *solar prominences.* Usually several of these red cloudlike formations are visible during a total eclipse. Some prominences actually erupt, speeding away from the Sun at close to a million miles per hour. (This movement is not evident to the naked eye during the few minutes of totality.) They arch above the surface and then disappear, sometimes lasting only a matter of hours. A few of these erupting prominences have been seen to reach a height of nearly one-third the diameter of the Sun itself.

This marvelous view of the Sun clearly commands the center of attention during totality. But there are other sights to see as well. Because the direct light of the Sun is blocked, some of the brighter stars and planets become visible. Sometimes a total solar eclipse reveals a small comet on its path near the Sun.

The darkness of totality resembles nighttime, and plants and animals react accordingly. Birds stop singing and may go to roost. Daytime flower blossoms begin to close as if for the night. Bees become disoriented and stop flying. The temperature drops in the coolness of the Moon's shadow. All of Nature seems still and quiet for this brief moment of daytime darkness.

Opposite: the diamond ring effect signals the beginning of totality for this eclipse photographer (February 26, 1979)

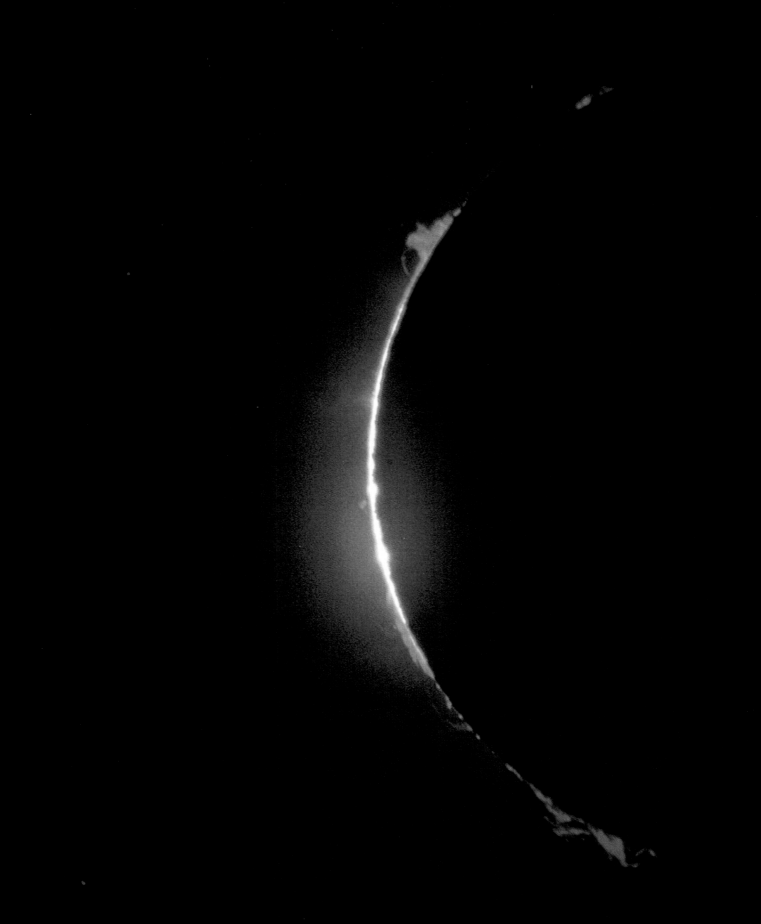

And then the shadow passes. A bright speck of sunlight flashes into view at the western edge of the Sun as the corona disappears. Totality has ended. The same events that preceded totality now occur in reverse order on the opposite side of the Sun. Baily's beads appear, followed by a thin crescent of sunlight. Daylight returns as more and more of the Sun is gradually uncovered by the passing Moon.

Finally the complete disk of the Sun is restored. The eclipse is over. The Moon continues in its orbit around the Earth, casting its shadow off into the vastness of space. Nothing tangible remains of the eclipse except some photographs and scientific data. Yet the memory of the experience is permanent—the fleeting beauty of the corona etched into the mind's eye by the sheer grandeur of the event. There is simply nothing else like it. And now it is gone—but not forever. The necessary alignment of Sun, Moon, and Earth will occur again to create other solar eclipses.

Engravings from the eclipse of August 7, 1869

Page 38: *Several solar prominences dot the eastern edge of the Sun at second contact—the moment totality begins. (March 7, 1970)*

Page 39: *The solar corona shines by light emitted from its ions as well as by sunlight reflected off electrons and dust particles. (February 16, 1980)*

Opposite: *Two skywatchers' peaks, separated by hundreds of years in history and thousands of miles on the Earth, share the Moon's shadow on the same day. The Pyramid of the Sun (top) at Teotihuacan, Mexico, and the new Keck Telescope (bottom) atop Mauna Kea, Hawaii, both lie within the path of totality on July 11, 1991.*

Patterns in Time and Space

In the 20th century, only nine total eclipse tracks cross the United States (excluding Alaska and Hawaii). The longest path of totality across the country came with the eclipse of June 8, 1918. This so-called "American eclipse" was observed from one corner of the nation to the other, all the way from Washington to Florida. Astronomers were joined by crowds of interested people along the eclipse track to witness the wonder of this day-time darkness.

The final total eclipse visible from the United States in this century is the eclipse of July 11, 1991, over Hawaii. The next one after that in the U.S. won't be until August 21, 2017. Of course, there will be total eclipses in between those dates; but none of them can be seen from any inhabited regions of North America.

The sequence of eclipses from year to year is determined by two different cycles of the Moon. The familiar monthly change of the phases of the Moon is one of these. The other cycle involves the gradual shift in orientation of the Moon's orbit. Only when these two cycles are favorably combined (about every six months) can a solar eclipse occur.

A solar eclipse may occur only at a new Moon. During this lunar phase the Moon passes between the Earth and the Sun. The Sun shines on the side of the Moon facing away from us, casting a shadow toward the Earth. A new Moon appears every 29½ days, but usually the Moon's shadow passes completely above or completely below the Earth. Since the Moon's orbit is tilted at a slight angle to the Earth's orbit, the Moon usually passes above or below the direct line of sight between the Earth and the Sun.

A *total solar eclipse* occurs when the umbra (the complete shadow of the Moon) sweeps across the Earth. During a *partial solar eclipse*, only the penumbra (the partial shadow of the Moon) touches our planet. The umbra passes either just above the North Pole or just below the South Pole, completely missing the Earth. No total eclipse is visible. Only partial phases (similar in appearance to the partial phases of a total eclipse) can be seen.

A third type of solar eclipse occurs when the Moon's umbra passes across the Earth, but is not quite long enough to touch the surface; the shadow cone diminishes to a point before reaching the Earth. This effect happens when the Moon is farther out in its orbit around the Earth. The Moon appears slightly smaller than the Sun and is not large enough to completely cover the Sun. When the Moon is centered over the Sun, a ring of sunlight remains visible around the edge. This type of eclipse is called an *annular eclipse*. (Annular comes from the Latin word meaning "ring.") Because the Sun is not completely covered by the Moon, the dramatic effects of a total eclipse (corona, darkness, etc.) are not present at either annular or partial eclipses of the Sun.

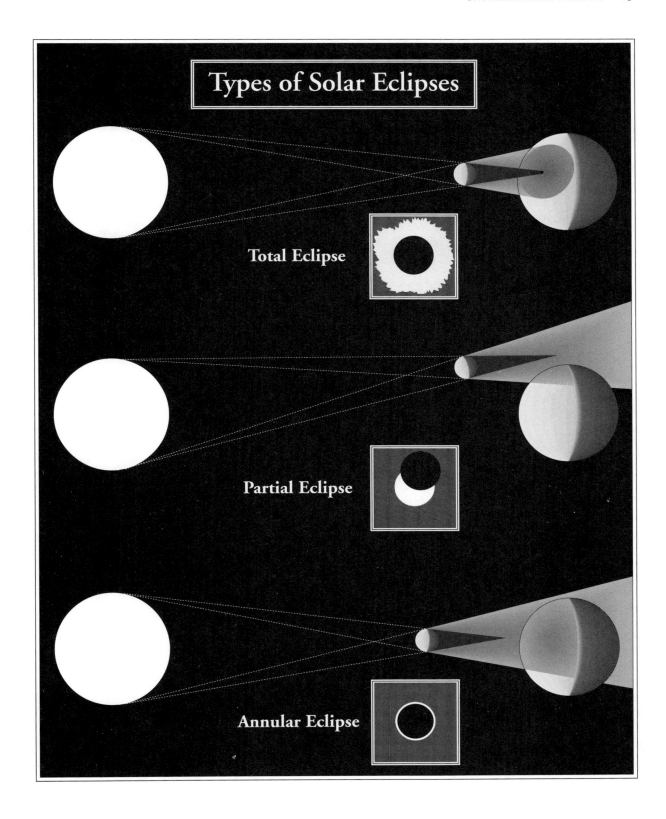

Types of Solar Eclipses

Total Eclipse

Partial Eclipse

Annular Eclipse

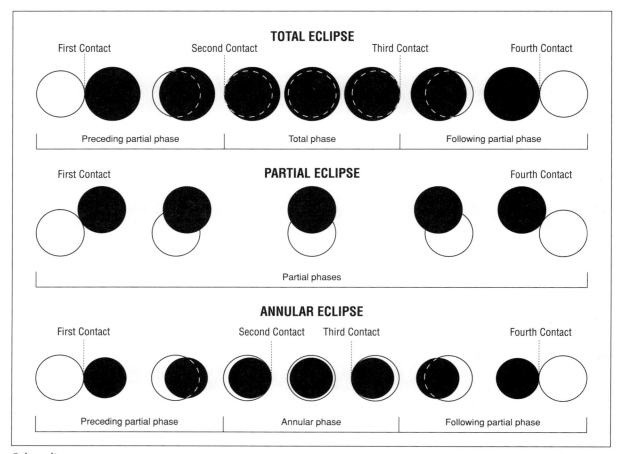

Solar eclipse contacts

Eclipse *contacts* (see diagram above) mark the transitions between different phases of a solar eclipse as viewed from a point on the Earth. For a **total eclipse**, *first contact* occurs at the instant the Sun's disk begins to be covered by the Moon. About an hour later, *second contact* marks the beginning of the total phase of the eclipse, and *third contact* signals the end of totality. The transition phenomena (Baily's beads, diamond ring, shadow bands, etc.) are visible near the second and third contacts. *Fourth contact* occurs when the Moon passes completely away from the disk of the Sun.

For a **partial eclipse** (or the partial phases of a total or annular eclipse), first contact also occurs at the instant when the Sun's disk begins to be covered by the Moon. But since there are only partial phases for this type of eclipse, there is no second or third contact. Fourth contact signals the end of the eclipse.

For an **annular eclipse**, second contact marks the point at which the disk of the Sun first completely surrounds the Moon, creating the ring of bright light around the Moon. Third contact occurs when the Moon moves on to break the ring of light of an annular eclipse.

Lunar Eclipses

An eclipse of the Moon, or lunar eclipse, takes place whenever the full Moon passes into the Earth's shadow, which is composed of two parts: the dark inner shadow, or *umbra*, and the lighter outer shadow, or *penumbra*. As in a solar eclipse, the Sun, Moon, and Earth are aligned in a straight line. But in this case, the Earth is between the Sun and the Moon. Also, there is not a lunar eclipse every month; usually the full Moon passes slightly above or below the Earth's shadow.

There are three types of lunar eclipses. A *total lunar eclipse* takes place when the Moon is completely engulfed by the Earth's shadow. This type of lunar eclipse offers the most unusual spectacle: the bright full Moon gradually darkens, and for up to an hour and a half appears a dull reddish or copper color. Although the Moon is in the complete shadow cone of the Earth, some sunlight is refracted by the Earth's atmosphere, with mostly the reddish wavelengths reaching the Moon. The exact color varies according to weather conditions at points near dawn and dusk at the time of the eclipse.

In a *partial lunar eclipse*, the umbra passes over only part of the Moon, causing only moderate darkening of the full Moon and little of the reddish color effect. During the time the umbra covers only part of the Moon (at either a partial or total lunar eclipse), the curved edge of the Earth's shadow, although somewhat fuzzy, can clearly be seen to be part of a circle. This projection of the Earth's circular shadow on the Moon is direct proof of the spherical shape of our planet. In a *penumbral lunar eclipse*, the Moon passes through only the penumbral portion of the Earth's shadow, and the darkening is scarcely noticeable.

Lunar eclipses occur at the rate of about 2.3 eclipses per year. Over time, total lunar eclipses account for about 36% of all lunar eclipses, partial about 27%, and penumbral about 37%. When a lunar eclipses does occur, it always happens 14¾ days before or after a solar eclipse. (Lunar eclipses occur during the eclipse seasons discussed on pages 50–51.) For example, there is a penumbral lunar eclipse on June 27, 1991, and another one on July 26, 1991, each occurring at the full Moon before and after the total solar eclipse of July 11, 1991.

Many more people witness total lunar eclipses than total solar eclipses, even though they occur with roughly the same frequency. (For example, in the period from 1901–2000, there are 81 total lunar eclipses and 70 total solar eclipses.) A lunar eclipse is visible from anywhere on the half of the Earth where it is nighttime during the hour or more that the Moon moves into the Earth's umbra (provided, of course, that the local night sky is clear enough to see the Moon). A total solar eclipse is visible only from within a narrow path of totality on a specific part of the Earth for just a few minutes at any location within that path.

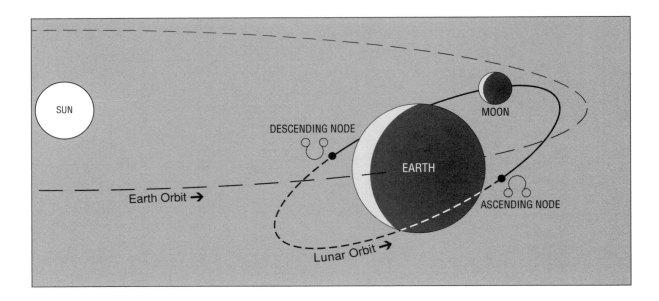

The Orbit of the Moon

As the Earth moves around the Sun in its yearly orbit, the direct line of sight between them sweeps through a plane called the *ecliptic*. The orbit of the Earth defines this plane. As seen from the Earth, the ecliptic is the path the Sun appears to take across the sky during the year. The twelve signs of the zodiac are distributed around this path.

The orbit of the Moon is tilted by about five degrees to the ecliptic. For half of its orbit, the Moon is above the ecliptic plane; the other half, below. The two points where the Moon's orbit intersects the ecliptic are called *nodes*. The *ascending node* marks the point of the Moon's passage to the upper part of its orbit; the *descending node* is the point where the Moon moves into its orbit below the ecliptic plane.

For a solar (or lunar) eclipse to happen, the Moon must be at or near one of its nodes when the new Moon (or full Moon) occurs. For a solar eclipse, the new Moon must occur when the Moon is close enough to the ecliptic plane so that the lunar shadow will cross some part of the Earth. This connection with eclipses is the reason the plane is named the ecliptic. A deeper, symbolic meaning is found in the astronomical symbols used for the Moon's ascending node (☊) and the descending node (☋). These symbols are generally supposed to represent the head and tail of the dragon swallowing the Sun according to the ancient belief about eclipses. In earlier times, the nodes were actually known by the fanciful titles "Dragon's Head" and "Dragon's Tail."

Opposite: Petrus Apianus' Astronomicum Cæsareum (1540) includes a rotating volvelle used to track the position of lunar nodes (depicted as the head and tail of a dragon)

F III

Corona seen on December 22, 1870

ꞇꞇ
Methinks it should be now a huge eclipse
Of Sun and Moon.
ꞇꞇ
Shakespeare, *Othello*

The Eclipse Year

The Moon and its orbit naturally move with the Earth as it travels around the Sun every year. If the lunar nodes were stationary with respect to the stars, the ascending node would be lined up between the Earth and the Sun at the same time each year (likewise for the descending node half a year later on the opposite side of the Earth's orbit). But the nodes of the lunar orbit are not quite stationary; they are gradually shifting their orientation in space. By the time a node is in line with the Sun again, it has regressed slightly. The alignment happens sooner than if the nodes were not moving. Thus it takes less than a full year for a node to be realigned between the Earth and the Sun. This period, called the *eclipse year*, is about 346.6 days long.

The diagram on the opposite page illustrates the regression of the Moon's nodes through an eclipse year. At position A the ascending node is lined up between the Earth and the Sun. This is the beginning of an eclipse year. As the Earth moves on (as in position B), the node passes out of the Earth–Sun alignment. About six months later position C is reached where the descending node lines up between the Earth and the Sun. But because the nodes themselves are slowly regressing, this alignment occurs a few days before the Earth reaches the point exactly opposite A. As the year continues, the descending node passes out of alignment (as in position D). Finally, at position E, the ascending node returns to a position between the Sun and the Earth. But because of the gradual regression of the nodes, the return takes only 346.6 days, an eclipse year. This is 18.6 days short of the full year it takes for the Earth to return to A.

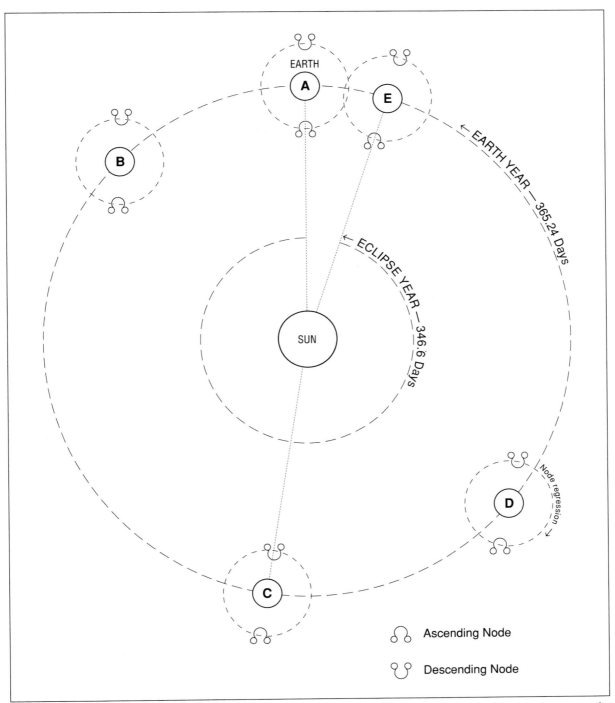

The eclipse year is marked by the return of a lunar node to the same position relative to the Sun and the Earth. Because the Moon's orbit (and the lunar nodes) are slowly regressing, it takes less than a full calendar year—346.6 days—to complete an eclipse year.

Eclipse Seasons

Recall that a solar eclipse occurs when the Moon passes between the Sun and the Earth (this is the definition of new Moon) at or near one of the lunar nodes. The new Moon need not perfectly coincide with a node to produce an eclipse. When the new Moon appears within 18¾ days before or after the alignment of a node, a solar eclipse will take place. This creates a 37½-day time window for eclipses. These periods when the conditions are favorable for an eclipse are called the *eclipse seasons*. These seasons occur whenever a node is near alignment (once every 173.3 days).

The diagram on the opposite page shows the timing of all the solar eclipses and eclipse seasons from 1990 to 2010. The eclipse seasons (shown as shaded horizontal bars) occur earlier and earlier each year. When a node returns to its alignment with the Sun, the calendar date will be about 19 days earlier than in the previous year. The repetition of these alignments gradually moves the eclipse seasons through all the months in a regular pattern over a period of years.

The diagram also reveals a pattern of repetition of new Moons in successive eclipse seasons at the same node. For example, at the descending node, the May 21, 1993, partial eclipse is followed by the May 10, 1994, annular eclipse. The May 10th eclipse falls on a date 11 days earlier than the eclipse of the previous year. The eclipses on April 29, 1995, and on April 17, 1996, follow the same pattern of dates separated by 11 (or sometimes 12) days. This happens because twelve new Moons separate each of these eclipses ($12 \times 29\frac{1}{2} = 354$ days). The 11¼-day difference between this time period and the calendar year accounts for this regular cycle of eclipse dates. But it can only last for several years: the new Moon on April 6, 1997, falls outside the next eclipse season at the descending node; the new Moon one month earlier (March 9, 1997) accounts for the solar eclipse during that season.

Because there is a new Moon every 29½ days, at least one new Moon appears during each 37½-day eclipse season. Therefore, at least one solar eclipse must occur during each eclipse season. Since there are at least two eclipse seasons every year, there must be at least two solar eclipses every calendar year. Some eclipse seasons have two solar eclipses, one near each end of the time window (e.g. July 1 and July 31 in the year 2000). Two "double seasons" in the same year produce a total of four solar eclipses.

The maximum number of solar eclipses in one calendar year is five, and this happens only rarely. In this case a year with two double seasons has the fifth eclipse squeezed in at the beginning of January or the end of December. This is possible because the eclipse year is shorter than a calendar year. This last occurred in 1935 when there were solar eclipses on January 5, February 3, June 30, July 30, and December 25. There won't be another calendar year containing five solar eclipses until the year 2206.

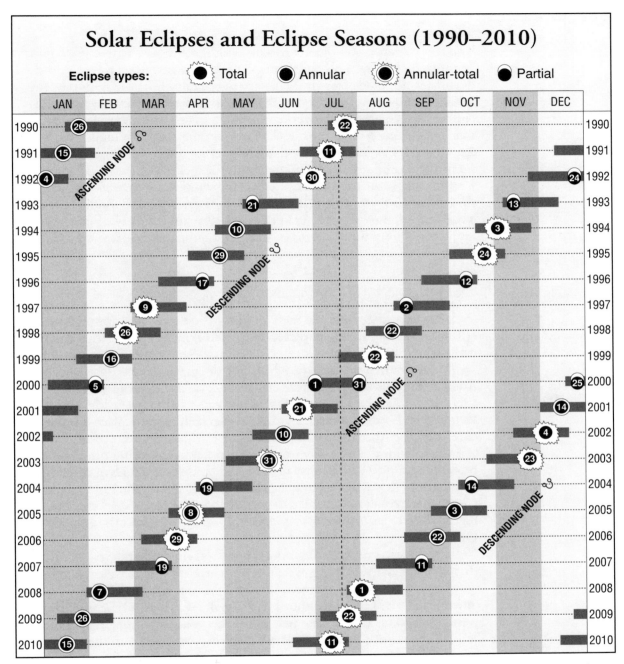

Solar eclipses can occur only during an eclipse season, a 37½-day window of time when a lunar node is in position between the Sun and the Earth. When a new Moon occurs near the midpoint of an eclipse season the result is an annular or total eclipse. A new Moon near the edge of the window produces a partial eclipse. The vertical dashed line in the center of the diagram connects the July 11, 1991, eclipse with the next eclipse (July 22, 2009) in the same saros series.

The *Saros* Cycle

The regular repetition of new Moons within successive eclipse seasons gets "out of synch" after only four years. That's because the eclipse year (346.62 days) does not come close to being an exact multiple of the 29½-day interval between new Moons (11 × 29½ = 324½ days, 12 × 29½ = 354 days). Astronomers have calculated the interval between each new Moon, called the *synodic month* from the Greek word for "meeting" or "conjunction," as 29.5306 days. A longer cycle, close to an exact multiple of the synodic month and eclipse year, would be useful for making eclipse predictions. Such a cycle is the *saros*, discovered by Babylonian astronomers in ancient times. The *saros* (meaning "repetition") lasts exactly 223 synodic months. That's a period of 18 years 11⅓ days (or 18 years 10⅓ days if five February 29ths fall within the period). The *saros* coincides closely with 19 eclipse years:

> 223 synodic months (29.5306 days) = 6,585.32 days
> 19 eclipse years   (346.6200 days) = 6,585.78 days

This resonance between the periods of these two cycles produces a repetition of eclipses in a remarkably short time. (In terms of astronomical cycles, 18 years is a short time!)

To illustrate how the *saros* works, see the diagram of eclipses seasons from 1990 to 2010 on the previous page. A total solar eclipse occurs on July 11, 1991, the path of totality crossing parts of Hawaii, Mexico, and Central and South America. Because a solar eclipse takes place on that date, we know that the Moon must be new and that a node must be near alignment with the Sun. Eighteen years and eleven days later the *saros* cycle repeats. Because 19 eclipse years have passed, the same node is near alignment. It is a new Moon again because exactly 223 synodic months have passed. This results in a total solar eclipse on July 22, 2009. The cycle is repeated, but this eclipse is visible from India, China, and the Pacific Ocean.

The *saros* has an extra 0.32 portion of a day included in its period (6,585.32 days). When the cycle repeats, the Earth will have rotated beyond its position at the former eclipse by this fraction of a day. The subsequent eclipse will be seen about a third of the way around the globe to the west. After three *saros* cycles, an eclipse takes place near the original longitude of the eclipse 54 years earlier. However, each eclipse in the series moves a little farther in the same direction toward one of the poles of the Earth. The July 22, 2009, eclipse track falls at slightly more northerly latitudes than the eclipse of July 11, 1991. This gradual shift in latitude occurs because the new Moon at each succeeding *saros* moves slightly with respect to the node. The half-day difference between 19 eclipse years and 223 synodic months (6,585.78 − 6,585.32 = 0.46 days) causes this change.

A series of eclipses, each separated by this 18-year 11⅓-day cycle, is called a *saros* series. Because the resonance between 19 eclipse years and the *saros* is not exact (0.46-day difference), a *saros* series cannot go on indefinitely. Eventually a series reaches a point when the eclipses are no longer visible; the umbra passes too far above or below the Earth to be seen. A single *saros* series spans over 1,200 years and includes between 68 and 75 solar eclipses. The *saros* series that includes the July 11, 1991, eclipse is shown on page 69.

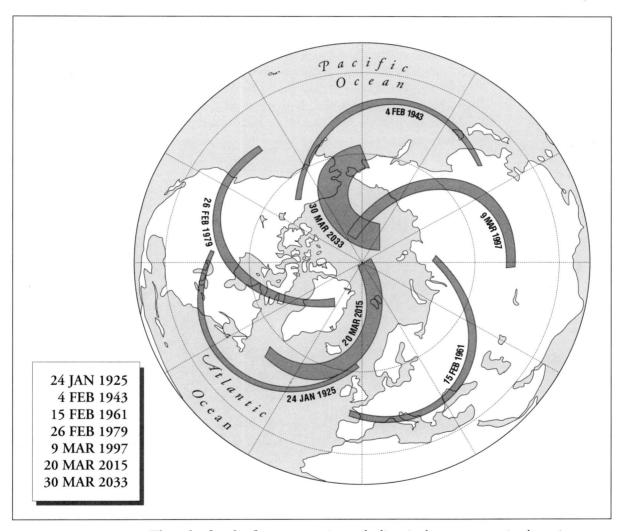

24 JAN 1925
4 FEB 1943
15 FEB 1961
26 FEB 1979
9 MAR 1997
20 MAR 2015
30 MAR 2033

The paths of totality for seven successive total eclipses in the same saros *series change in a regular pattern every 18 years. The paths, which are similar in shape, gradually widen and shift to more northerly latitudes. The longitude for each successive eclipse in the series shifts to the west a little more than one third of the way around the globe.*

Secrets of Stonehenge

There is no evidence that the builders of Stonehenge knew of the *saros*. But they didn't need to. The *saros* is a coincidence of celestial cycles; the marking of the monthly and yearly cycles at Stonehenge would work fine even if the *saros* did not exist. The likely purpose behind the pattern of these ancient stones and markers—to predict eclipses—has been revealed by the studies of two modern astronomers, Gerald Hawkins and Fred Hoyle.

The key to this fascinating discovery lies in the number 56. That is the number of so-called Aubrey holes (named after 17th century antiquarian John Aubrey) that form a ring about 300 feet in diameter around the large stone monuments in the center. Dug to a depth of several feet and then filled with chalk, these holes are evenly spaced around the circle. They were established about 2400 B.C. as part of the early phase of Stonehenge, some 300 years before the giant stone pillars were erected. The purpose of the Aubrey holes was for many years a mystery to those who studied Stonehenge.

Light-colored Aubrey holes (more evident at left) form a circle around Stonehenge

In his 1965 book *Stonehenge Decoded*, astronomer Gerald Hawkins explained his interpretation of the 56 Aubrey holes. Recall that the nodes of the Moon's orbit are regressing; they are moving slowly around the Earth from east to west. A complete revolution of these invisible orbital points takes 18.6 years (not to be confused with the 18-year *saros* cycle). To accurately predict eclipses from year to year, this regression cycle must be carefully recorded. Yet the builders of Stonehenge did not (as far as we know) use any form of writing. How could they mark this cycle? Three times 18.6 is very nearly 56. According to Hawkins, moving a marker (such as a small stone) three Aubrey holes each year makes a complete revolution of the system in 18⅔ years—close enough to the actual value for use in eclipse predictions. Two of these markers, each on opposite sides of the Aubrey ring, would show the position of the nodes at all times. The astronomical alignments at Stonehenge, too precise to be mere chance, imply a knowledge of the 18.6-year cycle of the lunar nodes. It seems plausible that the Aubrey holes were used to mark this cycle.

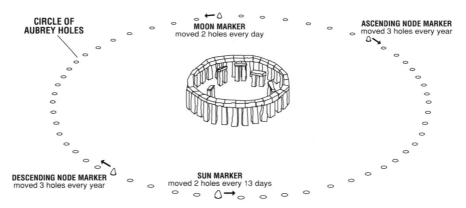

Fred Hoyle's 1977 book *On Stonehenge* shows how the number 56 may also be used to mark the movement of the Sun and the Moon. According to Hoyle, if a Sun marker is moved two holes every thirteen days, a complete circuit takes 364 days (56/2 × 13); this is only 1¼ days short of a full year. If a Moon marker is moved two holes every day, a complete circuit (28 days) is very close to the lunar month. These small discrepancies can be easily corrected using observations of the actual positions of the Sun and Moon.

Viewed in this way, Stonehenge represents a working model of the Sun–Moon–Earth system. The Sun, the Moon, and the lunar nodes, each represented by markers, revolve around the Earth located in the center. When the markers coincided, an eclipse would take place. All the necessary information (periodically corrected by actual observations) would have been available to the Stonehenge "astronomers" to use the 56 Aubrey holes as a primitive computer to predict eclipses.

Mysteries of Mexico

Halfway around the world from Stonehenge, another mysterious culture of skywatchers built stone monuments to mark the motion of celestial objects across the sky. The landscape of Mesoamerica is dotted with clusters of pyramids, temples, and observatories built by the Mayans and the Aztecs. These structures display a wealth of astronomical alignments, confirmed by research in the new discipline of archaeoastronomy, that are especially focused on the Sun, the Moon, the planet Venus, and the Pleiades star cluster. And unlike Stonehenge, surviving historical documents from the Mesoamerican period clearly indicate that these early cultures recorded eclipses and knew of eclipse cycles.

A key document showing knowledge of eclipse cycles is the Dresden Codex, one of four surviving Mayan hieroglyphic books that include astronomical tables and almanacs. The Dresden Codex (named for the German library that houses the document) contains black and red glyphs and vividly colored figures painted on bark paper made from the wild ficus tree. In addition to a series of 260-day almanacs and precise data on the motion of Venus, the Dresden Codex contains several pages of eclipse tables. These tables show groupings of specific glyphs: a series of glyphs for the numeral 177, followed by the glyph for the numeral 148, followed by a picture glyph for an eclipse. The numbers represent days, and the cumulative total of days recorded on several pages spans a period of more than 32 years.

In his 1980 book *Skywatchers of Ancient Mexico*, Anthony F. Aveni explains how these tables are related to eclipses. The numbers 177 and 148 are near exact multiples of 6 synodic months ($6 \times 29.53 = 177.18$ days) and 5 synodic months ($5 \times 29.53 = 147.65$ days). Once an eclipse is observed, these numbers can be used to count the days until the next "danger period" during which an eclipse may occur. Three intervals—**177 days**, **325 days** (177 + 148), and **354 days** (177 + 177)—may be used for either solar or lunar eclipses. For example, the list of notable future eclipses on page 92 of this book shows that the next total solar eclipse after July 11, 1991, occurs 355 days later on June 30, 1992. Next the list shows two 1994 solar eclipses (May 10 and November 3) separated by 177 days.

The eclipse table in the Dresden Codex even makes allowances for small corrections necessary over time. In a few instances, the number 178 appears in place of 177, indicating the need to add an extra day to the cycle. This would apply to the 1991–1992 interval discussed above (177 + 178 = 355 days). The expression of these complex cosmic cycles through a pair of numbers (177 and 148), along with the means to periodically correct the intervals based on observations, represents an amazing feat of understanding. One can only speculate what other Mesoamerican achievements would be worthy of our marvel today had more of their historical documents survived the onslaught of Western civilization.

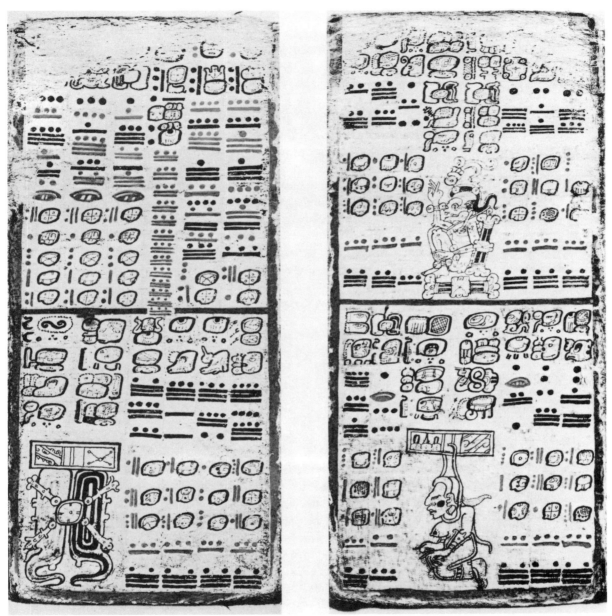

Pages 52 and 53 of the Dresden Codex show three large pictures, each representing an eclipse (one at the lower left and one each on the upper and lower panels on the right). The groups of horizontal lines and rows of dots are glyphs for numerals, some depicting 177- and 148-day eclipse intervals. Many of the surrounding glyphs stand for misery, malevolence, and death—all of which the Mayans supposedly associated with eclipses.

Motion of the Moon's Shadow

Stone monuments in various parts of the world show that ancient cultures had all the information needed to predict eclipses, including a way of keeping track of the nodes of the lunar orbit. The position of these invisible points in space—sometimes symbolized as dragons or serpents—helps determine the path of the Moon's shadow as it moves across the Earth toward the east during an eclipse.

Why does the shadow move eastward? Both the Moon and the Sun "rise" in the east and "set" in the west. This apparent motion across the sky (from east to west) is the result of the daily rotation of the Earth. Our planet is steadily spinning eastward; thus objects in the heavens (including the Sun and the Moon during an eclipse) seem to move toward the west. For a place in the path of totality, the entire duration of an eclipse (from first contact until fourth contact) is about two hours. During that time the Sun and Moon move through a part of their path across the sky from east to west.

The Moon's shadow, however, moves in the opposite direction: eastward. This happens because the Moon is revolving in its orbit from west to east. The umbra moves eastward with the Moon as it passes between the Sun and the Earth. This creates the effect of the Sun seeming to overtake the Moon during an eclipse. Because the Sun is farther away, it passes westward behind the Moon and casts a shadow that moves eastward on the Earth.

As the umbra sweeps eastward, an observer located in the path is also moving eastward due to the rotation of the Earth. But the Moon's shadow moves faster than any point on our rotating planet; the umbra always overtakes a stationary observer in its path. The faster the shadow is moving, the shorter the duration of totality. How fast the umbra moves over a point on the Earth's surface depends mainly on two factors: (1) the latitude (distance from the equator) of the point in the path, and (2) the time of day totality occurs at that point.

The first factor is the latitude. A point on the equator travels the complete circumference of the Earth (nearly 25,000 miles) in twenty-four hours, a rate of about 1,040 miles per hour. Points at higher latitudes (either farther north or farther south) don't have as far to go around the Earth in a day's rotation. The rotational speed of a point on the Earth becomes progressively slower at greater and greater distances from the equator. But the umbra's speed through space in the vicinity of the Earth is the same (about 2,100 miles per hour) regardless of latitude. The difference between this value and the speed of the point on the Earth helps determine the speed of the umbra across that point. The result is that the shadow moves more slowly in the tropics (within 23½ degrees latitude of the equator) and moves faster at points of greater latitude.

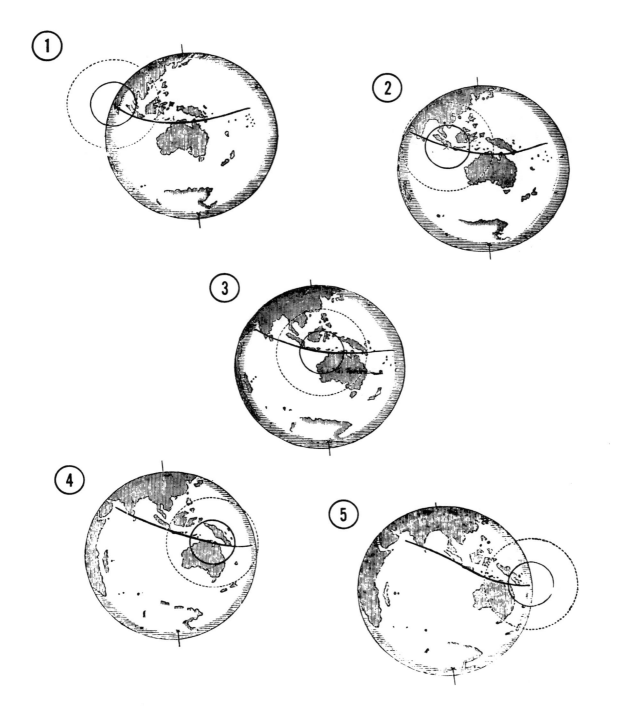

Motion of the Moon's shadow during the eclipse of December 12, 1871. Notice how the Earth rotated about 45 degrees from the start to the finish of the eclipse (about 3 hours).

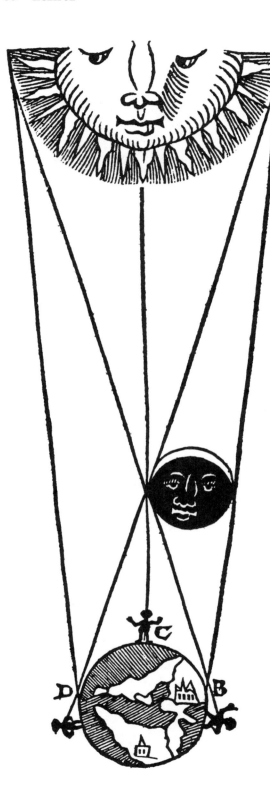

The other major factor affecting the speed of the shadow is the time of day the eclipse takes place. The umbra moves slowest across places where totality happens at noontime. When totality occurs earlier or later in the day, the umbra strikes the Earth at an oblique angle. The shadow at sunrise or sunset moves faster across the land or water than if it were closer to being perpendicular to the Earth's surface.

The combination of these two factors—latitude and time of day—determines the speed of the umbra. The umbra moves slowest when totality occurs at noon in the tropics. The speed is important because it affects the duration of the eclipse: the slower the shadow, the longer the time of totality. But another factor is just as important in determining the duration of totality: the width of the path made by the Moon's shadow.

Solar eclipse from Peurbach's Theoricæ Novæ Planetarum *(1553)*

Opposite: *James Ferguson's "Eclipsareon" from* Chamber's Encyclopædia *(London, 1779). This astronomical contrivance could exhibit the "time, quantity, duration, and progress of solar eclipses." The date and time of an eclipse would first be set on the dials at the base of the globe. Then the frame containing the screen of concentric circles would be adjusted for the Moon's latitude. A light source (such as a candle) would be used to project the shadow. By turning the crank, the circular screen would move across the frame, casting its simulated umbra on the rotating globe.*

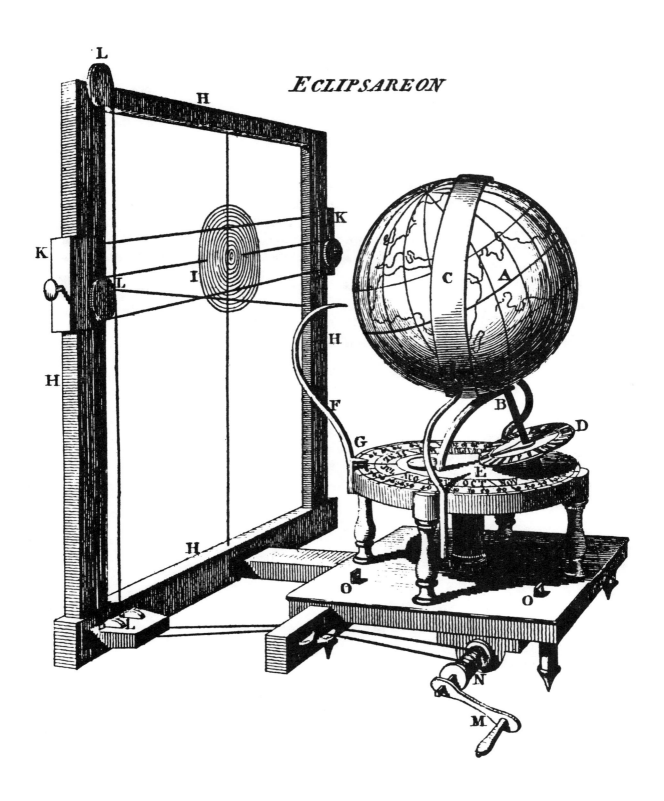

ECLIPSAREON

The Width of the Path

The width of the shadow at the Earth depends on how far away the Sun and the Moon are at the time of the eclipse. These distances vary because the orbits of the Earth and the Moon are not perfect circles, but ellipses. In an elliptical orbit, there are two extreme points: one where the orbiting body is closest to the body it revolves around, and another point where the orbiting body is farthest away. The umbra reaches its maximum diameter at the earth if an eclipse occurs when the Earth is farthest from the Sun, called *aphelion*, and when the Moon is closest to the Earth, called *perigee*. (These terms are formed with the prefixes *ap-* or *apo-* meaning "from," and *peri-* meaning "near"; *helion* refers to the Sun and *gee* to the Earth.)

Under these conditions the Sun appears slightly smaller and the Moon slightly larger than average. Thus the Sun can be covered by the Moon for a longer time. A wide umbra moving slowly across the Earth produces a long total eclipse. The longest possible duration of totality, 7 minutes and 31 seconds, occurs when the following conditions are met:

(1) The observer is at the equator
(2) Totality occurs there at noon
(3) The Earth is farthest from the Sun (aphelion)
(4) The Moon is closest to the Earth (perigee)

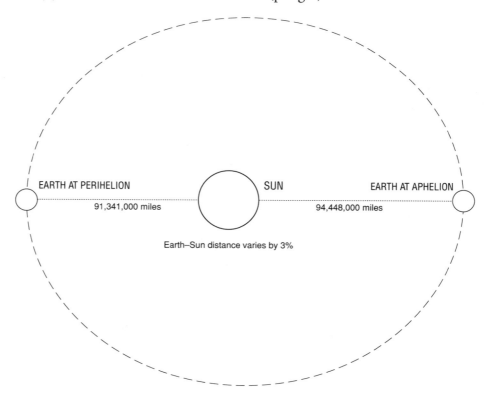

EARTH AT PERIHELION   SUN   EARTH AT APHELION

91,341,000 miles   94,448,000 miles

Earth–Sun distance varies by 3%

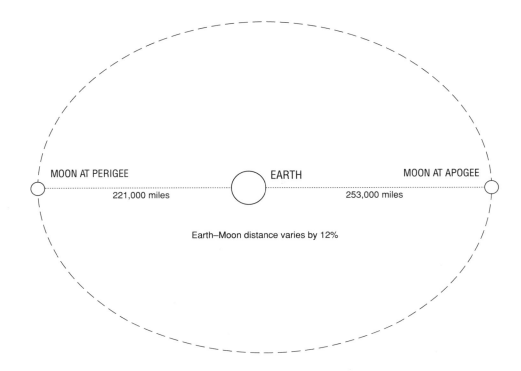

The total eclipse of June 20, 1955, the longest of this century, came close to meeting these conditions: it reached a maximum duration of 7 minutes and 7.7 seconds in the South China Sea. On July 16, 2186, the Sun will be eclipsed by the Moon for 7 minutes and 29 seconds, approaching the theoretical maximum.

The Earth–Sun distance and the Earth–Moon distance vary in regular cycles. The Earth orbits the Sun once a year, reaching its closest point (perihelion) in early January and its farthest point (aphelion) in early July. However, these orbital points are not stationary; each year they occur a fraction of a day later in time. Their slow revolution around the Sun takes about 20,000 years for a complete cycle. Ten thousand years ago the positions were reversed: perihelion was reached in July and aphelion in January. Ten thousand years from now this will again be the situation.

The Moon's orbit follows a similar motion around the Earth, but the time scale is greatly reduced. The Moon returns to its perigee (or apogee) in about two days less than it takes for a full Moon cycle (synodic month). This shorter 27½-day cycle is called the *anomalistic month*. This is the time it takes for the Moon to return to its original position the same distance from the Earth. These orbital points (lunar perigee and apogee) are slowly revolving around the Earth, making a complete circuit every 8.85 years.

Annular Eclipses

The Moon passes from apogee to perigee and back again every 27½ days; the Earth–Sun distance varies on a yearly cycle. If an eclipse occurs when the Moon is near apogee (Moon farther from Earth) and the Earth is nearer to perihelion (Sun closer to Earth), the Sun will appear larger than the Moon. The result is an *annular eclipse*, during which the Moon will not be able to completely cover the Sun, leaving a thin ring of sunlight visible around the dark lunar disk.

During an annular eclipse, the Moon's umbra falls short of reaching the Earth, producing what is called a *negative shadow*. This is an extension of the umbra projected onto the Earth. Anywhere within the path of this negative shadow the eclipse can be seen as annular, with the Sun completely surrounding the Moon from behind. Outside the negative shadow, within the penumbra, the eclipse appears as partial. When an annular eclipse takes place with the Earth at perihelion and the Moon at apogee, the negative shadow attains its greatest width, as much as 230 miles.

Most eclipses occur when the Sun and Moon are somewhere in between their closest and farthest points. If an eclipse is not a partial one, the relative effect of the distance is calculated to determine whether the type is annular or total.

A fourth type of eclipse has both annular and total phases. Called an *annular-total eclipse* (or sometimes a *central eclipse*), it starts out as annular, then becomes total, and finally reverts to annular, all in the same sweep of the shadow across the Earth. This rare type of eclipse occurs when the shadow cone of the umbra comes to a point right at the Earth. In the middle part of the eclipse, near noontime, the umbra just barely touches the Earth. The path of totality is very narrow. During the earlier and later phases of the eclipse, the umbra is not quite long enough to reach the points in the path around either side of the globe. Annular-total eclipses account for only about one in every 25 solar eclipses.

Annular eclipse and negative shadow

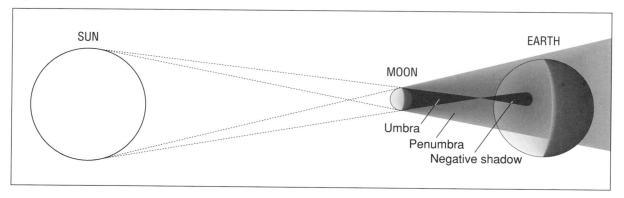

SUN

EARTH

MOON

Umbra

Penumbra

Negative shadow

Opposite: *Diagram of the solar eclipse of July 18, 1860*

VOL. IV.—No. 185.]   NEW YORK, SATURDAY, JULY 14, 1860.   [PRICE FIVE CENTS.

Entered according to Act of Congress, in the Year 1860, by Harper & Brothers, in the Clerk's Office of the District Court for the Southern District of New York.

ECLIPSE OF THE SUN ON THE EIGHTEENTH JULY.

On 18th July inst. an eclipse of the sun will take place, which will be more or less visible throughout the United States and Canada. We publish below a diagram of the eclipse. The reader must bear in mind that it represents the degree of observation at New York; hence at all places north of this parallel the eclipse will be greater, while at all places south of New York it will be less than is represented in the diagram.

It is hardly necessary to observe that an eclipse of the sun is caused by the passage of the moon between the earth and the sun. The motions of the heavenly bodies being governed by fixed mathematical laws, each eclipse can be predicted with certainty. The first appearance of the eclipse of 18th inst. since the creation of the world (according to sacred chronology) was in the year A.D. 958, December 8, old style, at 10 o'clock 50 minutes forenoon, when the moon's penumbra just came in contact with the earth at the south pole; it has appeared every nineteenth year since, and at

until the expiration of 12,492 years, when it will come on again at the south pole, and go through a similar course. The velocity of the moon's shadow across the earth during the eclipse will be about 1850 miles an hour, or four times the velocity of a cannon-ball.

DIAGRAM OF THE ECLIPSE OF THE SUN ON JULY 18, 1860.

tween the Indian Territory and New Mexico; it will then take a northeasterly and then a southeasterly course over the earth. The umbra, or total dark shadow of the moon, will first come in contact with the earth in the Pacific Ocean, one hundred miles west of the coast of Oregon, direct-

and Labrador to Cape Chidley, which will be the most favorable position on the Continent for observing the total eclipse. It will then enter the Atlantic Ocean, passing due east until nearly south of Cape Farewell, the southern cape of Greenland, where the sun will be totally eclipsed at noon of that place; it will then take a curved line toward the southeast, passing over the north of Spain, the Mediterranean Sea, Algiers, Tripoli, Fezzan, the southwestern corner of Egypt, into Nubia, where it will leave the earth near the Red Sea, a little before the setting of the sun at that place. The path of the umbra, in which the sun will be totally eclipsed, will be only about seventy miles in width; whereas the penumbra, in which the sun will appear more or less eclipsed, will extend from the Gulf of Mexico to 20 degrees upon the opposite side of the north pole, a distance of over six thousand miles. The umbra, in its passage over the earth, makes a curved line; this is caused by the spherical form of the earth. If the earth were a flat surface, the path of the umbra would then be a straight line from northwest to southeast, making an angle with the equator of 17 degrees. At all places south of the line of total eclipse the sun's northern limb will be eclipsed; but in Europe, England, Ireland, Greenland, Iceland, and the northern part of British America, the southern limb will be eclipsed.

each return the moon's shadow passed across the earth from west to east a little farther to the north at each return, until the year 1644, March 8, old style, when the centre of the moon's shadow passed a little to the north of the earth's centre (the moon being 14 minutes 46 seconds from her descending node, which was its 38th periodical return). It has continued to appear every nineteenth year since 1644, until this eclipse, which is its sixty-first periodical return. Its next appearance will be in 1878, July 29, at 3 o'clock 23 minutes in the morning, invisible in the United States. It will also appear again in 1896, August 9. It will continue to appear every nineteenth year until the year 2274, April 25, when the moon's shadow will just touch the earth at the north pole, which will be its seventy-sixth periodical and last appearance,

THE PATH OF THE ECLIPSE.

The penumbra, or partial shadow of the moon, will first come in contact with the earth at the rising of the sun in the northern part of Texas, be-

ly west of Oregon city, and a little to the southwest of the mouth of the Columbia River. It will then pass in a northeasterly direction over British America to Hudson's Bay, near Fort York, at the mouth of Nelson's River, crossing Hudson's Bay

HOW TO USE THE DIAGRAM.

The dark circular shade on the hemisphere represents the moon passing between the earth and sun; the shaded disk represents the sun partially eclipsed by the moon. To view this eclipse, as it will appear in the heavens, July 18, at 8 o'clock 10 minutes in the morning (New York time), face the east, take hold of the top of the diagram with your left hand, and the bottom with your right, with the back of the diagram toward the eye—incline the top of the diagram toward the north at an angle of about 45°, so that the north pole of the hemisphere will point as near as possible to the North Star; with the diagram in this position look through the back toward the sun at the time of the eclipse, and you will then have a true representation of the eclipse, and the exact position of the earth, moon, and sun at the time of the greatest obscuration, and the appearance it will present viewed through a smoked

Total Solar Eclipses (1900–1991)

The table on these two pages lists all the total and annular-total solar eclipses from 1900 to 1991. Each eclipse is described by its date, the maximum duration of totality, the maximum width of the path of totality, and the major geographical areas along the path from which totality is visible. The six eclipses shown in bold, all in the same *saros* series, represent the longest total solar eclipses of the twentieth century. (For a list of total solar eclipses from 1991 to 2035, see page 88.)

Date	Maximum Duration (min:sec)	Maximum Width (miles)	Path of Totality
1900 MAY 28	2:10	57	Mexico, United States, Spain, North Africa
1901 MAY 18	**6:28**	147	Indian Ocean, Sumatra, Borneo, New Guinea
1903 SEP 21	2:11	150	Indian Ocean, Antarctica
1904 SEP 9	6:19	145	Pacific Ocean
1905 AUG 30	3:46	119	Canada, Spain, North Africa, Arabia
1907 JAN 14	2:24	117	Russia, China, Mongolia
1908 JAN 3	4:13	92	Pacific Ocean
1908 DEC 23*	0:11	6	S. America, Atlantic Ocean, Indian Ocean
1909 JUN 17*	0:23	32	Greenland, Russia
1910 MAY 9	4:15	370	Antarctica
1911 APR 28	4:57	118	Pacific Ocean
1912 APR 17*	0:01	1	Atlantic Ocean, Europe, Russia
1912 OCT 10	1:55	52	Colombia, Brazil, Atlantic Ocean
1914 AUG 21	2:14	105	Greenland, Europe, Middle East
1916 FEB 3	2:36	67	Colombia, Venezuela, Atlantic Ocean
1918 JUN 8	2:22	70	Pacific Ocean, United States
1919 MAY 29	**6:50**	152	South America, Atlantic Ocean, Africa
1921 OCT 1	1:52	180	Antarctica
1922 SEP 21	5:58	140	Indian Ocean, Australia
1923 SEP 10	3:36	103	Pacific Ocean, Southern California, Mexico
1925 JAN 24	2:32	128	Great Lakes, NE United States, Atlantic Ocean
1926 JAN 14	4:10	91	Africa, Indian Ocean, Borneo
1927 JUN 29	0:50	48	England, Scandinavia, Arctic Ocean, Siberia
1928 MAY 19†			(Umbra barely touched Antarctica)
1929 MAY 9	5:06	120	Indian Ocean, Malaya, Philippines
1930 APR 28*	0:01	1	Pacific Ocean, United States, Canada
1930 OCT 21	1:55	52	South Pacific Ocean
1932 AUG 31	1:44	96	Arctic Ocean, Canada, New England
1934 FEB 14	2:52	76	Borneo, Pacific Ocean
1936 JUN 19	2:31	82	Greece, Turkey, Soviet Union, Japan

* Annular-total eclipse

† Last total eclipse in *saros* series

Date	Maximum Duration (min:sec)	Maximum Width (miles)	Path of Totality
1937 JUN 8	**7:04**	155	Pacific Ocean, Peru
1938 MAY 29‡	4:04	420	South Atlantic Ocean
1939 OCT 12	1:32	260	Antarctica
1940 OCT 1	5:35	135	Colombia, Brazil, Atlantic Ocean, S. Africa
1941 SEP 21	3:21	88	Soviet Union, China, Pacific Ocean
1943 FEB 4	2:39	142	Japan, Pacific Ocean, Alaska
1944 JAN 25	4:08	90	Peru, Brazil, Western Africa
1945 JUL 9	1:15	57	U.S., Can., Greenland, Scandinavia, U.S.S.R.
1947 MAY 20	5:13	121	South America, Atlantic Ocean, Africa
1948 NOV 1	1:55	52	Africa, Indian Ocean
1950 SEP 12	1:13	83	Arctic Ocean, Siberia, Pacific Ocean
1952 FEB 25	3:09	85	Africa, Middle East, Soviet Union
1954 JUN 30	2:35	95	U.S., Canada, Iceland, Europe, Middle East
1955 JUN 20	**7:07**	157	Southeast Asia, Philippines, Pacific Ocean
1956 JUN 8	4:44	266	South Pacific Ocean
1957 OCT 23†			(Umbra barely touched Antarctica)
1958 OCT 12	5:10	129	Pacific Ocean, Chile, Argentina
1959 OCT 2	3:01	75	New England, Atlantic Ocean, Africa
1961 FEB 15	2:45	160	Europe, Soviet Union
1962 FEB 5	4:08	91	Borneo, New Guinea, Pacific Ocean
1963 JUL 20	1:39	63	Pacific Ocean, Alaska, Canada, Maine
1965 MAY 30	5:15	123	New Zealand, Pacific Ocean
1966 NOV 12	1:57	52	Pacific Ocean, South America, Atlantic Ocean
1967 NOV 2‡			(Umbra barely touched Antarctica)
1968 SEP 22	0:39	64	Soviet Union, China
1970 MAR 7	3:27	95	Pacific Ocean, Mexico, Eastern U.S., Canada
1972 JUL 10	2:35	109	Siberia, Alaska, Canada
1973 JUN 30	**7:03**	159	Atlantic Ocean, Central Africa, Indian Ocean
1974 JUN 20	5:08	214	Indian Ocean, Australia
1976 OCT 23	4:46	123	Africa, Indian Ocean, Australia
1977 OCT 12	2:37	61	Pacific Ocean, Colombia, Venezuela
1979 FEB 26	2:49	185	NW United States, Canada, Greenland
1980 FEB 16	4:08	92	Africa, Indian Ocean, India, Burma, China
1981 JUL 31	2:02	67	Soviet Union, Pacific Ocean
1983 JUN 11	5:10	123	Indian Ocean, Indonesia, New Guinea
1984 NOV 22	1:59	53	New Guinea, Pacific Ocean
1985 NOV 12	1:58	430	Antarctica
1986 OCT 3*	0:01	1	North Atlantic Ocean
1987 MAR 29*	0:07	3	South Atlantic Ocean, Africa
1988 MAR 18	3:46	104	Sumatra, Borneo, Philippines, Pacific Ocean
1990 JUL 22	2:32	125	Finland, Soviet Union, Aleutian Islands
1991 JUL 11	**6:53**	160	Hawaii, Mexico, C. America, Colombia, Brazil

* Annular-total eclipse

† Last total eclipse in *saros* series

‡ First total eclipse in *saros* series

Repetition of Eclipses

Where a total solar eclipse is visible depends mainly on three factors. First, the longitude of the eclipse path is determined by which part of the Earth happens to be in daylight when the Moon passes between the Earth and the Sun. Second, the latitude of the eclipse path is a function both of the season of the year (which part of the Earth is tilted toward the Sun) and of how close the Moon is to a node when the eclipse occurs. The path of totality sweeps near tropical latitudes for eclipses that occur when the Moon is near a node. And third, the area covered by totality depends on the width (and length) of the path. The wider the path, the more locations experience totality.

The repetition of eclipses in time follows a definite pattern. The *saros* is the best example of an eclipse cycle. Recall that after 18 years and 11⅓ days, a new Moon and a lunar node return almost exactly to their former alignment, and the eclipse is repeated. What's more, both eclipses are likely to be of the same type. This is a result of the close coincidence between the *saros* and 239 anomalistic months ($239 \times 27.5545 = 6{,}585.54$ days). When the *saros* repeats, the Moon will be almost the same distance from the Earth as before. Also, because the eclipses take place at the same time of the year (only 11 days' difference), the Earth–Sun distance is almost the same. This almost exact resonance between the eclipse year, the synodic month, and the anomalistic month, coupled with the *saros* being close to an even number of years, results in remarkable cycles centuries long.

The progression of eclipses through a complete *saros* series takes some 1,200 years. All the eclipses in a given series occur at either the ascending or descending node. If the eclipses take place at the ascending node, each path will fall a little farther south with each successive eclipse; if at the descending node, the eclipse tracks progress northward. (Some *saros* series experience occasional interruptions to this general trend due to the seasonal tilt of the Earth's axis.)

The *saros* series that includes the July 11, 1991, eclipse began six centuries ago at the descending node. On June 14, 1360, the Moon's penumbra barely grazed Antarctica. This partial eclipse was repeated on June 25, 1378, but this time more of the Earth was covered by the partial shadow. Each time the *saros* repeated, the Moon's shadow fell a little farther north. On September 8, 1504, the first annular eclipse of the series occurred. A series of annular and annular-total eclipses continued every 18 years until January 17, 1703, the date of the first total eclipse in this *saros* series. The total eclipses continue until the last one on May 13, 2496. After that, partial eclipses finish the series. The 70th and final eclipse of this *saros* series will be visible only as a partial eclipse from Arctic regions on July 19, 2604.

A Complete *Saros* Series of Solar Eclipses

Date	Type	Date	Type
1360 JUN 14	Partial	1991 JUL 11	Total
1378 JUN 25	Partial	2009 JUL 22	Total
1396 JUL 5	Partial	2027 AUG 2	Total
1414 JUL 17	Partial	2045 AUG 12	Total
1432 JUL 27	Partial	2063 AUG 24	Total
1450 AUG 7	Partial	2081 SEP 3	Total
1468 AUG 18	Partial	2099 SEP 14	Total
1486 AUG 29	Partial	2117 SEP 26	Total
1504 SEP 8	Annular	2135 OCT 7	Total
1522 SEP 19	Annular	2153 OCT 17	Total
1540 SEP 30	Annular	2171 OCT 29	Total
1558 OCT 11	Annular	2189 NOV 8	Total
1576 OCT 21	Annular	2207 NOV 20	Total
*1594 NOV 12	Annular	2225 DEC 1	Total
1612 NOV 22	Annular-total	2243 DEC 12	Total
1630 DEC 4	Annular-total	2261 DEC 22	Total
1648 DEC 14	Annular-total	2280 JAN 3	Total
1666 DEC 25	Annular-total	2298 JAN 13	Total
1685 JAN 5	Annular-total	2316 JAN 25	Total
1703 JAN 17	Total	2334 FEB 5	Total
1721 JAN 27	Total	2352 FEB 16	Total
1739 FEB 8	Total	2370 FEB 27	Total
1757 FEB 18	Total	2388 MAR 9	Total
1775 MAR 1	Total	2406 MAR 20	Total
1793 MAR 12	Total	2424 MAR 31	Total
1811 MAR 24	Total	2442 APR 11	Total
1829 APR 3	Total	2460 APR 21	Total
1847 APR 14	Total	2478 MAY 3	Total
1865 APR 25	Total	2496 MAY 13	Total
1883 MAY 6	Total	2514 MAY 25	Partial
1901 MAY 18	Total	2532 JUN 5	Partial
1919 MAY 29	Total	2550 JUN 16	Partial
1937 JUN 8	Total	2568 JUN 26	Partial
1955 JUN 20	Total	2586 JUL 7	Partial
1973 JUN 30	Total	2604 JUL 19	Partial

*Begin dates from Gregorian calendar

The repetition of eclipses follows very regular patterns in time. Eclipse seasons and *saros* cycles come and go like clockwork. Of the four types of solar eclipses, partial eclipses are most common. The following table gives values for the relative distribution of each type.

Type of Solar Eclipse	Proportion of All Solar Eclipses
Partial	35%
Annular	33%
Total	28%
Annular-total	4%

But the main attraction of eclipses is totality. The repetition of total solar eclipses at a given place on the Earth, however, does not seem to follow any discernible cycle. The map on the next page shows all total eclipse paths across North America since 1900. Goldendale, Washington, for example, is in the path of totality for the eclipses in 1918 and 1979. Yet many areas have not experienced a total eclipse for several centuries.

Partial phases of solar eclipses can be seen about every 2½ years from any particular spot on the Earth. The best estimate for total eclipses is to say they recur at the same location about every 360 years on the average. This figure is based on the average area of the paths of totality, the total surface area of the Earth, and the overall frequency of total eclipses. But the actual circumstances for particular locales vary, sometimes widely, from this estimate. The table below helps illustrate the apparent random nature of the recurrence of eclipses at the same place. The examples were chosen, not to prove any lack of pattern, but to present the flavor of the variation involved.

Location	Dates of Consecutive Total Eclipses	Years in Interval
London	29 OCT 878 A.D. — 22 APR 1715 A.D.	837
Jerusalem	30 SEP 1131 B.C. — 4 JUL 336 B.C.	795
Great Pyramid of Egypt	1 APR 2471 B.C. — 29 JUN 2159 B.C.	312
Honolulu, Hawaii	3 SEP 1690 A.D. — 7 AUG 1850 A.D.	160
Stonehenge	8 MAY 1169 B.C. — 7 MAY 1066 B.C.	103
Goldendale, Washington	9 JUN 1918 A.D. — 26 FEB 1979 A.D.	61
Mazatlán, Mexico	11 JUL 1991 A.D. — 8 APR 2024 A.D.	33
Yellowstone National Park	29 JUL 1878 A.D. — 1 JAN 1889 A.D.	11
Tomb of Tutankhamun	31 MAY 957 B.C. — 22 MAY 948 B.C.	9
Lake Okechobee, Florida	19 AUG 2259 A.D. — 22 DEC 2261 A.D.	2½
Southern New Guinea	11 JUN 1983 A.D. — 22 NOV 1984 A.D.	1½

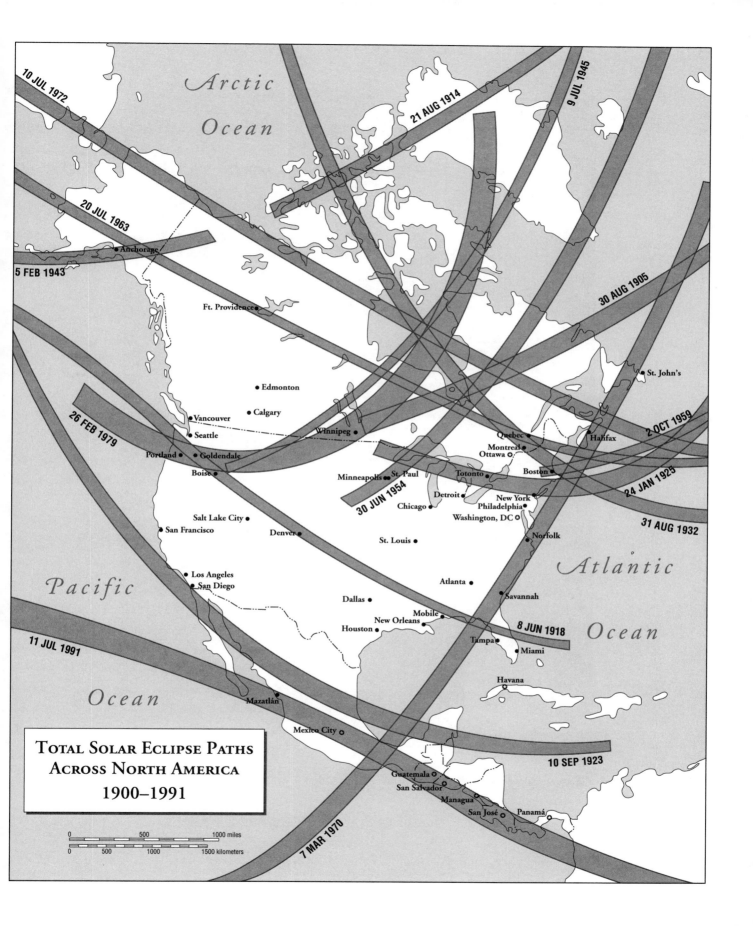

TOTAL SOLAR ECLIPSE PATHS
ACROSS NORTH AMERICA
1900–1991

CHAPTER 3

How To Observe an Eclipse

The spectacular sight of a total solar eclipse is for most of us a once-in-a-lifetime event. Unless you're an astronomer or an avid eclipse follower, you'll probably get only one chance to see it. It's estimated that only one in a thousand people ever experiences totality. This wondrous spectacle of the complete halo around the Sun can't be seen under any other earthly circumstances. Astronomers are able to observe part of the corona without an eclipse using a *coronagraph*, a kind of telescope invented in 1931. But the naked-eye view of the corona is visible only from within the path of totality.

In addition to the sight of the corona, there are other marvelous phenomena to observe during a total eclipse. The daytime darkness and the swift onset of the Moon's shadow add to the drama of the few short minutes the corona is visible. Shadow bands, Baily's beads, the reaction of plants and animals—all add to the excitement and impact of the inexorable alignment of Sun, Moon, and Earth. It's simply a matter of being in the right place at the right time—and knowing what to look for.

The time and location of each eclipse, of course, is different. But the observation site considerations and the viewing techniques are essentially the same for all total solar eclipses. The eclipse of July 11, 1991, is used as the example in this chapter to illustrate how to observe an eclipse.

Opposite: Path of the Moon's shadow on the Earth during the "New Year's Day Eclipse" of January 1, 1889. The illustration shows a direct telegraph line from San Francisco to New York that was made available to the astronomers in California for the purpose of sending word of the success of their eclipse observations to observers stationed in the path farther east.

The Path of Totality

Early in the morning of July 11, 1991, as seen from a point in the Pacific Ocean about 1,400 miles southwest of Hawaii, the Sun rises. But instead of the familiar orange ball coming up over the horizon, the blackened disk of the eclipsed Sun appears. The umbra, the complete shadow of the Moon, makes its first grazing contact with the edge of the Earth at sunrise and the total eclipse begins. The shadow races eastward at thousands of miles per hour toward Hawaii, where the Sun is already up this Thursday morning.

At 7:28 a.m., the umbra first touches land at Kailua Kona on the "Big Island" of Hawaii. The shadow, moving at more than 7,600 miles per hour, sweeps along a 138-mile-wide path, blotting out the Sun for up to 4 minutes and 8 seconds over the entire Big Island and barely grazing the southern coast of Maui. As the eclipse continues across the Pacific, the shadow slows its movement and the path gets slightly wider. It next reaches land in Mexico, where it crosses the tip of Baja California and moves inland on the western coast of Mexico at about noon. The event reaches greatest eclipse at a point just inland from the coast and north of Tepic. Here the shadow reaches its maximum width (160 miles) and totality reaches its maximum duration (6 minutes and 53.5 seconds).

By now the eclipse is about half over. At its slowest point, the shadow is moving about 1,400 miles per hour. It begins picking up speed as it moves southeast over Mexico and Central America. The path narrows slightly as it crosses five Latin American capitals: Mexico City, Guatemala, San Salvador, Managua, and San José. Late in the afternoon the shadow picks up more speed as it races across Colombia and northwestern Brazil. Finally, the total eclipse ends at sunset about 200 miles northwest of Brasília. The Moon's shadow leaves the Earth and continues its course through space.

The passage of the Moon's umbra over the Earth, from sunrise in the Pacific to sunset in Brazil, takes almost three and a half hours for this eclipse. During that time it travels about 9,300 miles across the surface of the Earth. That's an average speed of almost 3,000 miles per hour. The area covered by the path of totality amounts to approximately 1,360,000 square miles, or about 0.7 percent of the total surface area of the planet. Not only do millions of people live in the path, but the shadow passes over major vacation spots as well as the astronomical observatories on Mauna Kea. The partial phases of the eclipse are visible from almost all of the United States (except Alaska), much of Canada, all of Central America and the Caribbean, and a large portion of South America.

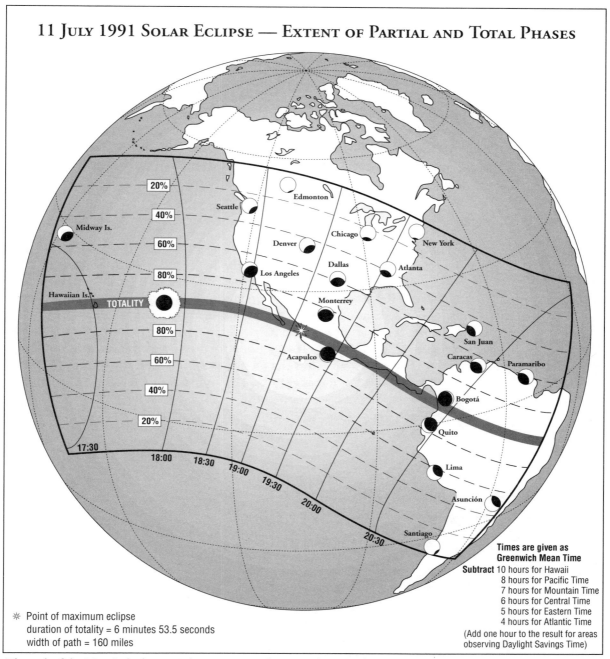

11 July 1991 Solar Eclipse — Extent of Partial and Total Phases

20%
40%
60%
80%

Midway Is.

Hawaiian Is.

TOTALITY

80%
60%
40%
20%

17:30
18:00
18:30
19:00
19:30
20:00
20:30

Seattle
Edmonton
Denver
Chicago
Dallas
New York
Los Angeles
Atlanta
Monterrey
Acapulco
San Juan
Caracas
Paramaribo
Bogotá
Quito
Lima
Asunción
Santiago

Times are given as Greenwich Mean Time
Subtract 10 hours for Hawaii
8 hours for Pacific Time
7 hours for Mountain Time
6 hours for Central Time
5 hours for Eastern Time
4 hours for Atlantic Time
(Add one hour to the result for areas observing Daylight Savings Time)

✳ Point of maximum eclipse
duration of totality = 6 minutes 53.5 seconds
width of path = 160 miles

The path of the Moon's shadow on July 11, 1991, is shown on the map above at 30-minute intervals as the shadow sweeps from west to east. The dashed lines running east-west show curves of equal magnitude of the partial phases at 20% increments, indicating the greatest percentage of the Sun covered by the Moon as seen from points along these lines. The farther away from the path of totality, the less of the Sun's disk is blocked out. (For more detail on the path of totality, see the map on pages 78-79.)

Selecting an Observation Site

The partial phases of a total solar eclipse are usually visible over a wide area (most of the Western Hemisphere on July 11, 1991). But only within the path of totality can you see the spectacular and striking effects. The difference between experiencing a total eclipse and a partial eclipse is, literally, "the difference between night and day." Those who live within the path or take the opportunity to travel there have the chance to be rewarded with one of the most fleeting and beautiful visions of Nature's grandeur.

Your choice of a site within the path should be guided by three main factors:

(1) Duration of totality at the site
(2) Unobstructed view of the Sun
(3) Chances for clear skies

The map on pages 78-79 shows the July 11, 1991 path of totality over populous areas, from Hawaii to Brazil. For any given locale, a point nearer the central line of the eclipse experiences a longer totality; this is because the Moon's shadow, which forms an ellipse on the surface of the Earth, is wider nearer the center of the path. The accompanying diagram shows this variation across the path. If you are located just within the path, totality will not last very long—less than a minute. However, the "edge phenomena" of a total eclipse (Baily's beads, diamond-ring effect, and view of the chromosphere) will last longer there.

Width of the shadow and duration of totality decrease farther away from central line of eclipse

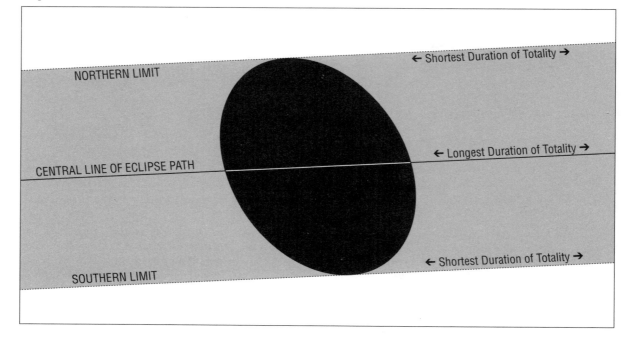

To ensure an unobstructed view of the eclipse, you need to know approximately where the Sun will be in the sky. You don't want any trees or mountains, for example, blocking your view. An easy way to find this out is to look for the Sun from your vantage point a day or two before the eclipse at the same time of day totality occurs. Or you can use the map on the next two pages to determine the Sun's position at total eclipse.

For the July 11, 1991, eclipse, the Sun is low in the sky and toward the direction east northeast in Hawaii; in Mexico, where the eclipse occurs closer to noon, the Sun is higher, more directly overhead; nearer Colombia, the eclipsed Sun will appear lower in the western sky. The compass direction (azimuth) of the Sun is shown for selected locations along the path.

The Sun's angle above the true horizon (altitude) is also shown. It's important to realize that the angle of altitude (for example, 21° on Hawaii) is measured from the true or level horizon. Also, if you're on a hill or tall building with a good view of the west, you may get a chance to see the approach of the Moon's shadow as it races toward you.

❧

Roses have thorns, and silver fountains mud;
Clouds and eclipses stain both Moon and Sun.

❧

Shakespeare, *Sonnet XXXV*

The third factor in choosing a site is the weather. Unless you fly above the clouds to observe an eclipse, you'll always have to take some risk on the weather. But there is a lot you can do to optimize your chances to see the eclipse. The meteorological term for cloudy skies is "sky cover." It is measured in numbers from 0 to 10, with each number representing a tenth. (For example, a sky cover of 3 indicates 30% overcast, or 70% clear.) Weather Bureau records will show the average sky cover for different places along the path at the time of year of the eclipse.

The weather prospects for the July 11, 1991, eclipse vary widely along the path. On the Big Island, the Kohala Coast and Mauna Kea will have a 70% or 80% chance for clear skies. Baja California also has good chances for clear skies. Points in the Mexican interior and along the Pacific coast of Central America will more likely be cloudy for this event.

These sky cover predictions are only general estimates covering large areas. Local weather conditions can be very different for places a short distance apart. You'll want to avoid places likely to have fog; also, stay away from mountain ridges where clouds tend to gather. But perhaps the greatest asset in finding clear skies for an eclipse is mobility. Driving a few miles to a clearer location at the last minute could save the day for you.

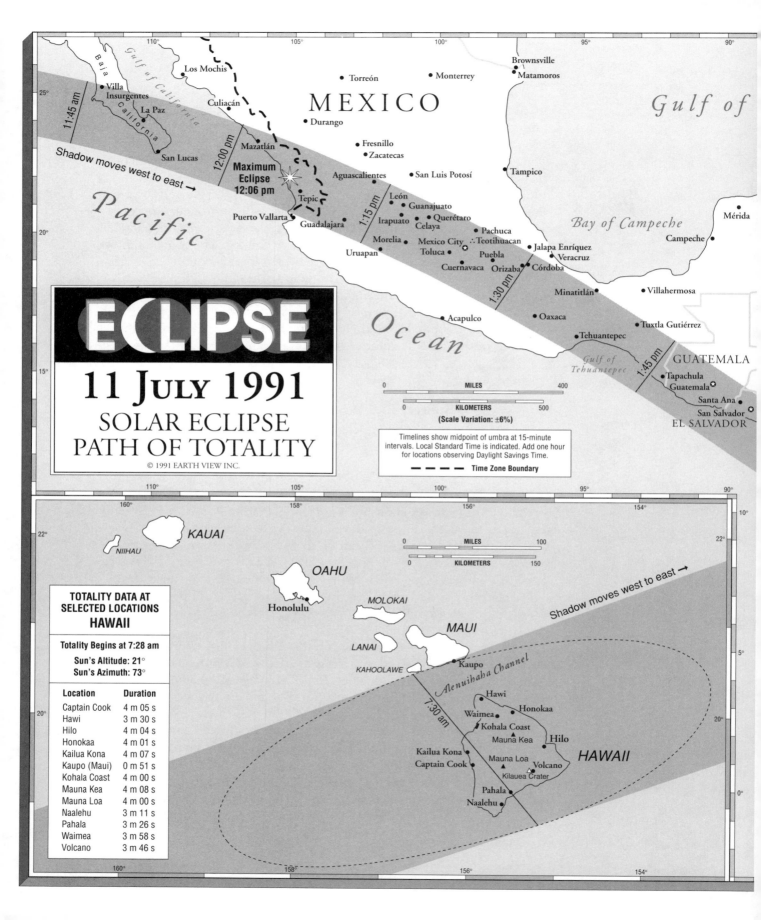

ECLIPSE
11 JULY 1991
SOLAR ECLIPSE PATH OF TOTALITY
© 1991 EARTH VIEW INC.

Map labels (Mexico / Central America panel):

110° · 105° · 100° · 95° · 90°

Baja California · Gulf of California

Los Mochis · Torreón · Monterrey · Brownsville · Matamoros

Villa Insurgentes · Culiacán · MEXICO · Gulf of

La Paz · Durango

San Lucas · Mazatlán · Fresnillo · Zacatecas · Tampico

11:45 am · 12:00 pm · Maximum Eclipse 12:06 pm · Aguascalientes · San Luis Potosí

Shadow moves west to east →

Tepic · León · Guanajuato · Bay of Campeche · Mérida

Pacific · Puerto Vallarta · Guadalajara · Irapuato · Querétaro · Celaya · Pachuca · Campeche

1:15 pm · Morelia · Mexico City · Teotihuacan · Jalapa Enríquez · Veracruz

Uruapan · Toluca · Puebla · Orizaba · Córdoba

Ocean · Cuernavaca · Minatitlán · Villahermosa

Acapulco · Oaxaca · 1:30 pm

Tehuantepec · Tuxtla Gutiérrez

Gulf of Tehuantepec · 1:45 pm · GUATEMALA

Tapachula · Guatemala

Santa Ana

San Salvador · EL SALVADOR

Scale box:

MILES 0 — 400
KILOMETERS 0 — 500
(Scale Variation: ±6%)

Timelines show midpoint of umbra at 15-minute intervals. Local Standard Time is indicated. Add one hour for locations observing Daylight Savings Time.

– – – – Time Zone Boundary

Hawaii panel labels:

160° · 158° · 156° · 154°

22° · KAUAI · NIIHAU

OAHU · MILES 0 — 100 · KILOMETERS 0 — 150

Honolulu · MOLOKAI · Shadow moves west to east →

MAUI · LANAI · KAHOOLAWE · Kaupo · Alenuihaha Channel

7:30 am · Hawi · Waimea · Honokaa · Kohala Coast · Mauna Kea · Hilo

Kailua Kona · Mauna Loa · HAWAII · Captain Cook · Volcano · Kilauea Crater

Pahala · Naalehu

TOTALITY DATA AT SELECTED LOCATIONS
HAWAII

Totality Begins at 7:28 am

Sun's Altitude: 21°
Sun's Azimuth: 73°

Location	Duration
Captain Cook	4 m 05 s
Hawi	3 m 30 s
Hilo	4 m 04 s
Honokaa	4 m 01 s
Kailua Kona	4 m 07 s
Kaupo (Maui)	0 m 51 s
Kohala Coast	4 m 00 s
Mauna Kea	4 m 08 s
Mauna Loa	4 m 00 s
Naalehu	3 m 11 s
Pahala	3 m 26 s
Waimea	3 m 58 s
Volcano	3 m 46 s

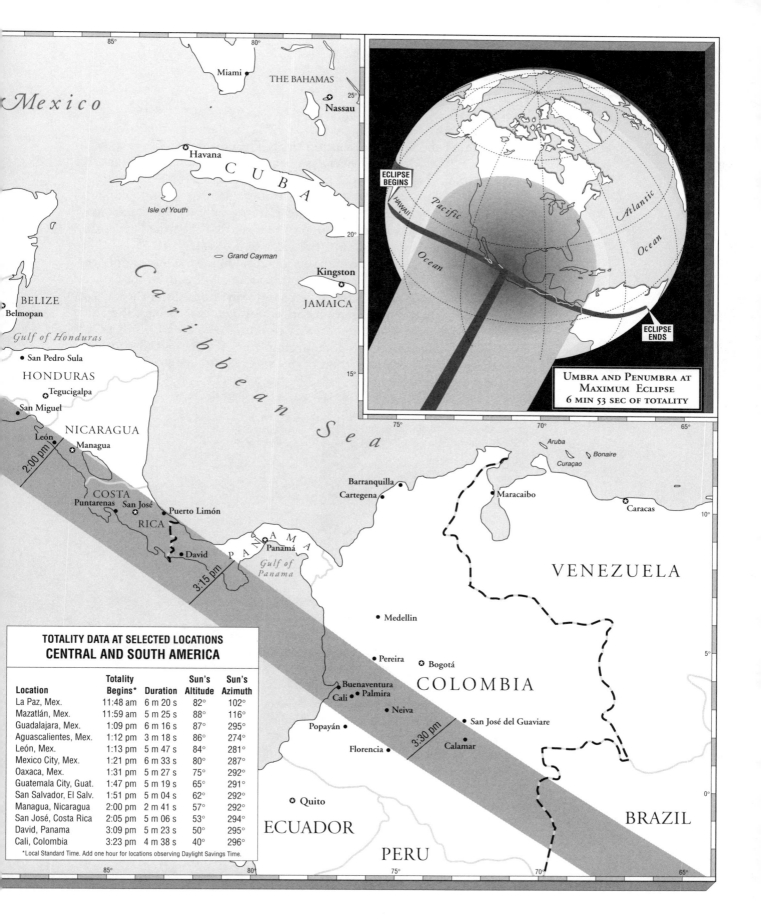

Miami

THE BAHAMAS

Nassau

25°

Mexico

20°

Havana

C U B A

Isle of Youth

Grand Cayman

Kingston

JAMAICA

BELIZE
Belmopan

Gulf of Honduras

San Pedro Sula

HONDURAS

Tegucigalpa

San Miguel

León

NICARAGUA

Managua

2:00 pm

COSTA
RICA

Puntarenas

San José

Puerto Limón

David

P A N A M A

Panamá

Gulf of Panama

3:15 pm

C a r i b b e a n   S e a

15°

10°

Barranquilla

Cartegena

Aruba

Bonaire

Curaçao

Maracaibo

Caracas

VENEZUELA

Medellin

Pereira

Bogotá

COLOMBIA

Buenaventura

Cali

Palmira

Neiva

Popayán

San José del Guaviare

Florencia

Calamar

3:30 pm

5°

0°

Quito

ECUADOR

PERU

BRAZIL

ECLIPSE BEGINS

HAWAII

Pacific

Ocean

Atlantic

Ocean

ECLIPSE ENDS

UMBRA AND PENUMBRA AT
MAXIMUM ECLIPSE
6 MIN 53 SEC OF TOTALITY

TOTALITY DATA AT SELECTED LOCATIONS
CENTRAL AND SOUTH AMERICA

Location	Totality Begins*	Duration	Sun's Altitude	Sun's Azimuth
La Paz, Mex.	11:48 am	6 m 20 s	82°	102°
Mazatlán, Mex.	11:59 am	5 m 25 s	88°	116°
Guadalajara, Mex.	1:09 pm	6 m 16 s	87°	295°
Aguascalientes, Mex.	1:12 pm	3 m 18 s	86°	274°
León, Mex.	1:13 pm	5 m 47 s	84°	281°
Mexico City, Mex.	1:21 pm	6 m 33 s	80°	287°
Oaxaca, Mex.	1:31 pm	5 m 27 s	75°	292°
Guatemala City, Guat.	1:47 pm	5 m 19 s	65°	291°
San Salvador, El Salv.	1:51 pm	5 m 04 s	62°	292°
Managua, Nicaragua	2:00 pm	2 m 41 s	57°	292°
San José, Costa Rica	2:05 pm	5 m 06 s	53°	294°
David, Panama	3:09 pm	5 m 23 s	50°	295°
Cali, Colombia	3:23 pm	4 m 38 s	40°	296°

*Local Standard Time. Add one hour for locations observing Daylight Savings Time.

85°   80°

85°   80°   75°   70°   65°

Observation Safety Precautions

The total phase of a solar eclipse, when the sky is dark and the corona is visible around the Sun, is a beautiful sight. The best way to observe the event during these few brief minutes is simply to look directly at this glimmering halo in the sky. The corona is a million times fainter than the bright disk of the Sun; there is no danger of eye damage when looking directly at the corona or the prominences during totality. Binoculars may reveal even finer detail, but most observers agree that the naked eye is the best "instrument" for viewing the full glory of the event.

For about an hour before and after the total phase the Sun is only partially obscured. This is when it is dangerous to look directly at the Sun. Normally the Sun is too bright to look at anyway. But during the partial phases (especially when only a thin crescent of light is showing), the Sun does not appear as bright, and you may be tempted to look directly at it. DON'T DO IT! The danger of damaging your eyes does not depend on brightness. As long as any portion of the Sun's disk remains visible it can still cause eye damage.

Total Phase: safe to view directly

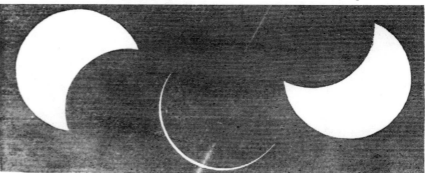

Partial Phases: DANGEROUS TO LOOK DIRECTLY AT THE SUN, even if only a thin crescent is visible

The lenses of your eyes act as tiny magnifiers; if you look at the partially eclipsed Sun, its rays are focused on the retina of your eyes and can burn them. This is the same sort of thing that happens when you use a magnifying glass to focus the Sun to a pinpoint on paper or leaves to burn a hole in them. The only difference is that it is your eyes that would be burned. Part of the danger lies in the fact that the retina is not sensitive to pain; you wouldn't even feel it happening. But a retinal burn is permanent and irreversible, producing a blank spot in the most vital part of your field of vision.

Astronomers observe the sun directly through professionally manufactured optical filters that screen out the hazardous rays of the Sun. But unless you are trained in their use, it is not recommended that you try this method. And you're taking a big chance if you try to improvise your own filter. During the March 7, 1970, eclipse in the United States there were 145 reported cases of people who damaged their eyes by looking at the partially eclipsed Sun either directly or through sunglasses, exposed film, smoked glass, and the like. None of these homemade devices can be guaranteed safe. Play it smart and don't take any chances with your precious gift of vision.

Using smoked glass or other homemade filters, as did these observers for an eclipse in 1865, is NOT RECOM-MENDED for viewing the partial phases of a solar eclipse. Serious eye damage could result.

Safe Viewing Techniques

There are some perfectly safe ways to observe the partial phases of the eclipse without looking directly at the Sun. These methods involve viewing the image of the Sun projected onto some surface; the image can be focused by having the sunlight pass through a pinhole. This is the same effect seen when the light from the partially eclipsed Sun shines through the leaves of a tree, creating tiny crescent images in the shadow on the ground. The diagram below illustrates how to build and use a simple pinhole projector. This is the safe and recommended way to observe the passage of the Moon across the face of the Sun during the partial phases of a solar eclipse. (If the eclipse is televised, it would also be safe to view it on the TV screen.) And don't forget: during the few minutes of totality it's OK to look directly at the Sun's corona.

Building and using a pinhole projector to view the partial phases of a solar eclipse

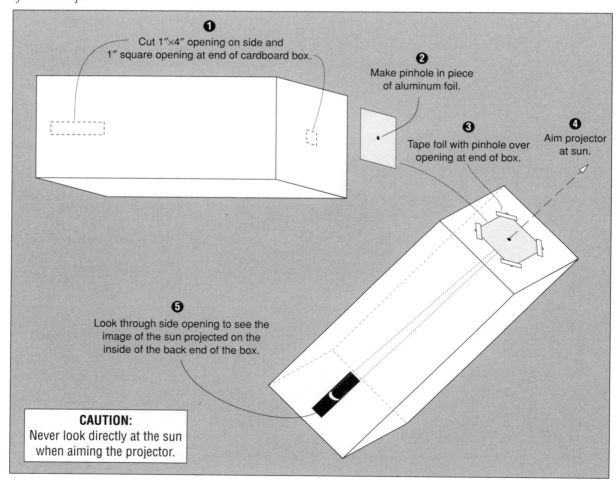

❶ Cut 1″×4″ opening on side and 1″ square opening at end of cardboard box.

❷ Make pinhole in piece of aluminum foil.

❸ Tape foil with pinhole over opening at end of box.

❹ Aim projector at sun.

❺ Look through side opening to see the image of the sun projected on the inside of the back end of the box.

CAUTION:
Never look directly at the sun when aiming the projector.

You might find it interesting to see how you would judge the degree of darkness during totality. Scientists at eclipses in the past, before sensitive light-measuring instruments were available, carried out elaborate experiments to obtain some measure of the darkness. Comparisons were made to candlelight, moonlit nights, twilight, etc. Reports were given on the readability of instrument dials and various sizes of print. One experimenter even proposed that the opening and closing of plant leaves and blossoms be used as a gauge of the relative darkness in the Moon's shadow. In general, it is darker nearer the center of the path of totality and in clearer weather. (Clouds scatter light.) You may also want to observe shadow bands. Put up a flat white sheet or screen at least three feet wide facing the Sun and look closely for the faint ripples of light a few minutes just before and just after totality.

It is perfectly safe to look directly at the Sun's corona, as did these observers on July 18, 1860, during the few minutes of the total phase of a solar eclipse.

Capturing Solar Eclipse Images

Taking pictures of a total eclipse can be as simple as aiming your pocket camera and pushing the button, or as complex as using sophisticated cameras mounted on telescopes driven by motors. You're not likely to get good results with the first method, but you needn't go all the way to the other extreme to produce some satisfying pictures of the corona during totality. The whole field of photography is filled with technical details. The discussion here is intended only as an overview—enough basic information to let you decide if you want to get more details from some of the sources listed as references at the end of the book.

The main problem in using a small pocket camera to photograph the corona is the short focal length of the lens; pictures taken with these types of cameras produce a very small image of the black disk of the eclipse. A camera with a lens of greater focal length will produce better results. For example, a 600mm lens will produce a circular image about ¼ inch in diameter. This is a good fit for 35mm film and gives you some leeway for errors in centering the image. It's a good idea to mount your camera on a tripod and to bracket exposure times from 1/500 second to 2 seconds; for exposures longer than 2 seconds for this size lens you should use an equatorial drive mount. (This device is explained in astro-photography references.) Choice of film seems to be a matter of personal preference; something in the range from ASA 50 to ASA 200 should be adequate. And there is no need for any camera filters during totality.

To shoot the partial phases of the eclipse you'll need to use a filter or two to produce the equivalent of a 5.00 neutral density filter. Be sure you don't try to look at the Sun through these filters; they are designed for photographic use only and are not safe for your eyes. And don't look through the viewfinder at the partially eclipsed Sun. A good subject for a camera lens of shorter focal length is a multiple exposure of the complete sequence of the eclipse from first to fourth contact. Use the filters for exposures of the partial phases every 5 or 6 minutes, and take one exposure of totality with the filters removed. Be sure that your camera is securely mounted and that you don't knock it out of position during the two hours or so you have it set up.

There are some good photo subjects during an eclipse other than the Sun itself. You may want to try to capture shadow bands on high-speed film using short exposures. But don't feel too disappointed if they elude your camera; they have proven very difficult to photograph, and they aren't visible at every eclipse. The crescent images of the partially eclipsed Sun seen in the shadow of a tree can make a good picture. Or you may want to try a sequence of shots showing the darkness of the landscape before, during, and after totality.

If you're interested in solar eclipse photography, check the references for more detailed sources of information. Or you may want to get in touch with a local amateur astronomy group; there you can swap ideas with people who have learned from experience. But whatever your attempts to photograph an eclipse, don't get so lost in your camera that you forget to look up at the corona—a sight whose beauty no film can reveal nearly as well as the human eye itself.

Because the light of the Sun is blocked out during totality, the sky turns dark and stars and planets become visible. Although this daytime darkness lasts only a few minutes, many of the other objects in the heavens may be seen and identified. The accompanying sky maps show which stars and planets are in the sky at the eclipse on July 11, 1991. The Sun is in the constellation Gemini. Venus, Mars, Mercury, and Jupiter are in the vicinity of the eclipsed Sun on that day, but will be visible only from Latin America where the Sun is higher in the sky. On rare occasions (as in 1882 in Egypt) a small comet may be visible.

Stars and planets in the vicinity of the eclipsed Sun on July 11, 1991 as seen from Baja California. The Sun appears almost directly overhead in Mexico during totality.

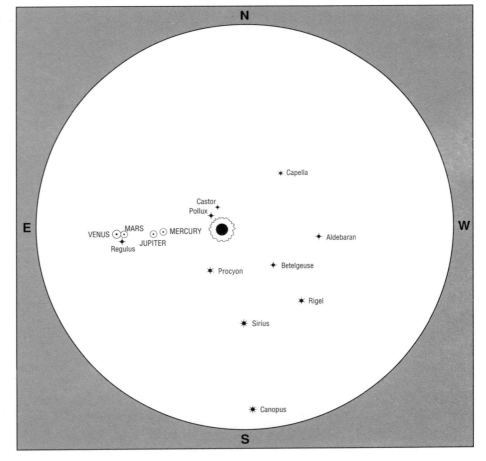

The advent of the camcorder has made videotaping solar eclipses a popular activity, especially since most consumer models now use technology that is impervious to overexposure. (Aiming earlier tube-type videocameras at the Sun can severely damage the equipment.) A typical camcorder (with ½-inch CCD sensor and a 66mm lens) will produce an image of the lunar disk about one inch in diameter on a 13-inch diagonal TV screen. (The corona will appear about three inches in diameter.) Use a solar filter for the partial phases, and then remove it for the period of totality. You can use the full Moon, which is roughly equivalent in brightness to the light from the corona, to test your camera beforehand. At the time of the eclipse, you may also want to record the onset of the shadow and the sunset effect around the horizon. Again, don't forget to look up from your camera and enjoy the eyewitness beauty of the event.

Observing totality is a way of experiencing not just these brief events but a larger sense of our solar system as well. Being in the path is a unique way of becoming part of this perfect alignment of the Sun, the Moon, and the Earth. Few ever forget the experience of totality.

Stars in the vicinity of the eclipsed Sun on July, 11, 1991 as seen from Hawaii. The sun appears 21° above the horizon near ENE during totality in Hawaii. The planets visible later from Mexico are still below the horizon in Hawaii.

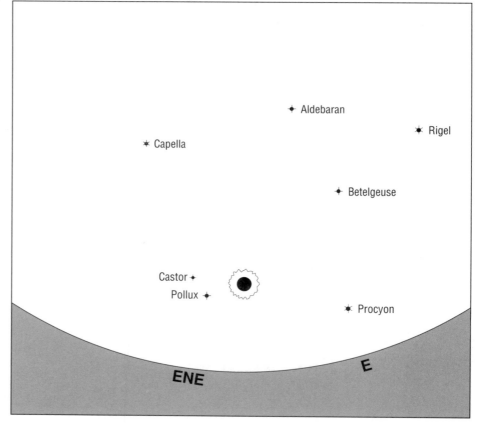

Observation Checklist

Here is a checklist for observing an eclipse. Site selection criteria and observation equipment are summarized, followed by a list of phenomena in complete sequence for a total solar eclipse.

SITE SELECTION

☐ Site near center of path   ☐ Unobstructed view of Sun

_____ **Duration of totality**   _____ Sun's altitude (angle above true horizon)

     _____ Sun's azimuth (or compass direction)

Eclipse times   _____ Sky cover prediction

_____ First Contact (eclipse begins)

_____ Second Contact (totality begins)   ☐ **Provisions for last-minute mobility**

_____ Third Contact (totality ends)

_____ Fourth Contact (eclipse ends)

EQUIPMENT

☐ Pinhole projector   ☐ Binoculars   ☐ Cameras, film, etc.

(USE BINOCULARS ONLY DURING TOTALITY)

ECLIPSE EVENTS

☐ **First Contact (eclipse begins)**   ☐ **Third Contact (totality ends)**
☐ Moon begins to cover Sun   ☐ Darkness passes
☐ Crescent images of Sun   ☐ Diamond ring effect
☐ Gradual darkening   ☐ Baily's beads
☐ Shadow bands (2 or 3 minutes before totality)   ☐ Shadow bands
☐ Baily's beads (½ minute before totality)   ☐ Gradual lightening of sky
☐ Diamond ring effect   ☐ Crescent images of Sun
☐ Approach of shadow from the west   ☐ Gradual uncovering of Sun

☐ **Second Contact (totality begins)**   ☐ **Fourth Contact (eclipse ends)**
☐ Corona
☐ Prominences   Notes:
☐ Chromosphere
☐ Stars and planets
☐ Darkness of landscape
☐ Plant/animal reactions
☐ Temperature drop

Total Solar Eclipses (1991–2035)

Date	Maximum Duration (min:sec)	Maximum Width (miles)	Path of Totality
1991 JUL 11	**6:53**	160	Hawaii, Mexico, Central America, Colombia, Brazil
1992 JUN 30	5:20	182	South Atlantic Ocean
1994 NOV 3	4:23	117	Peru, Bolivia, Paraguay, Brazil
1995 OCT 24	2:09	48	Iran, India, Southeast Asia
1997 MAR 9	2:50	221	Mongolia, Siberia
1998 FEB 26	4:08	94	Galapagos Islands, Panama, Colombia, Venezuela, Guadeloupe, Montserrat, Antigua
1999 AUG 11	2:22	69	Europe, Middle East, India
2001 JUN 21	4:56	124	Atlantic Ocean, Southern Africa
2002 DEC 4	2:03	54	Southern Africa, Indian Ocean, Australia
2003 NOV 23	1:57	308	Antarctica
2005 APR 8*	0:42	16	South Pacific Ocean
2006 MAR 29	4:06	114	Africa, Turkey, Soviet Union
2008 AUG 1	2:27	147	Greenland, Soviet Union, China
2009 JUL 22	**6:38**	160	India, China, Pacific Ocean
2010 JUL 11	5:20	160	S. Pacific Ocean, southern tip of S. America
2012 NOV 13	4:02	111	Australia, Pacific Ocean
2013 NOV 3*	1:39	35	Atlantic Ocean, central Africa
2015 MAR 20	2:46	287	N. Atlantic Ocean, Norwegian Sea, Svalbard
2016 MAR 9	4:09	96	Indonesia, North Pacific Ocean
2017 AUG 21	2:40	71	U. S. (from Oregon to South Carolina)
2019 JUL 2	4:32	124	South Pacific Ocean, Chile, Argentina
2020 DEC 14	2:09	56	Chile, Argentina
2021 DEC 4	1:54	260	Antarctica
2023 APR 20*	1:16	30	Indonesia
2024 APR 8	4:28	122	Mexico, United States, Canada
2026 AUG 12	2:18	182	Greenland, Iceland, Spain
2027 AUG 2	**6:22**	160	Gibraltar, North Africa, Saudi Arabia
2028 JUL 22	5:09	143	Indian Ocean, Australia, New Zealand
2030 NOV 25	3:43	105	S. Africa, Indian Ocean, Australia
2031 NOV 14*	1:08	24	Pacific Ocean
2033 MAR 30	2:37	483	Alaska, Arctic Ocean
2034 MAR 20	4:09	99	Central Africa, Middle East, China
2035 SEP 2	2:54	72	China, North Korea, Japan, Pacific Ocean

*Annular-total eclipse

Eclipses shown in bold are in the same *saros* series that includes July 11, 1991

EPILOGUE

The Future

July 11, 1991, is the last time in the 20th century that the path of totality crosses any part of North America. The next total solar eclipse visible from the United States won't occur until August 21, 2017. In the intervening 26 years there are 16 total solar eclipses, but they can be seen only from other parts of the world.

Twenty-six years without a total eclipse over the entire North American continent is unusual. But there are also times of plenty. In the 38 years between 1594 and 1632 there were eight total eclipses visible from the U.S. or Canada. And in the 23rd century, six total eclipses will be visible from the United States in the years 2245, 2252, 2254, 2259, 2261, and 2263—a busy 18 years.

Total eclipses will continue to occur on the Earth at the predictable rate of about once every year and a half. The 43-year period from 1992 to 2035 provides 32 chances to witness totality. (This includes total and annular-total eclipses.) If you were able to position yourself on the Earth at the point of maximum eclipse for each of these events, *and* you were blessed with clear skies during totality, you could spend 107 minutes in the shadow of the Moon. That's an average of about 2½ minutes per year under optimum conditions. The extreme rarity of totality, coupled with its visual beauty and true cosmic grandeur, make it one of the most exquisite natural experiences available on planet Earth.

Our Sun and Moon have planned a number of interesting opportunities to experience totality in the next four decades. Some paths cross highly populated areas. Others touch the Earth in exotic vacation spots. Still others only graze remote ocean regions. At each appointed rendezvous, the Moon's shadow is sure to be there. Perhaps you will be there, too.

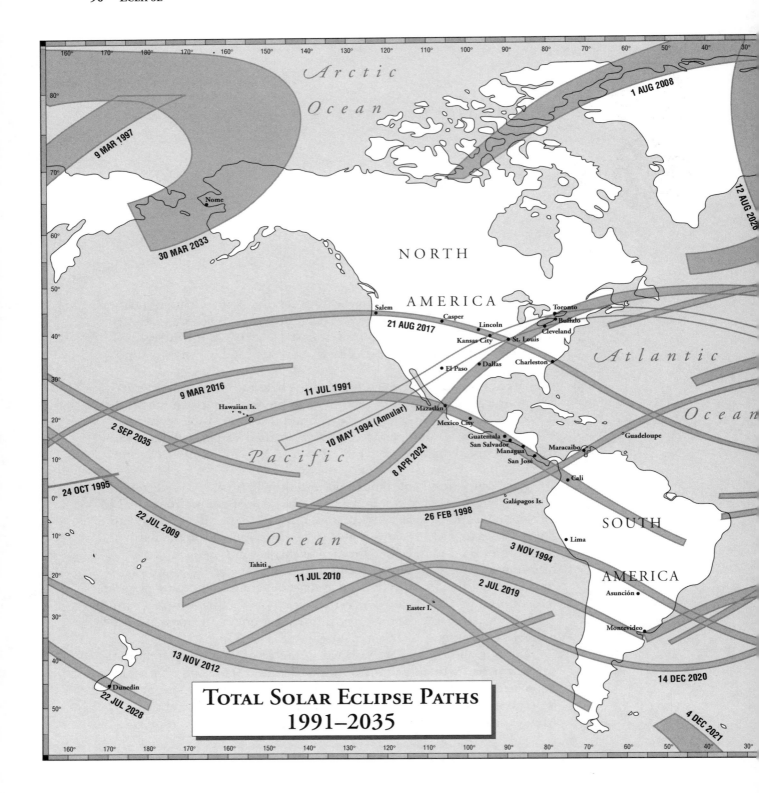

TOTAL SOLAR ECLIPSE PATHS 1991–2035

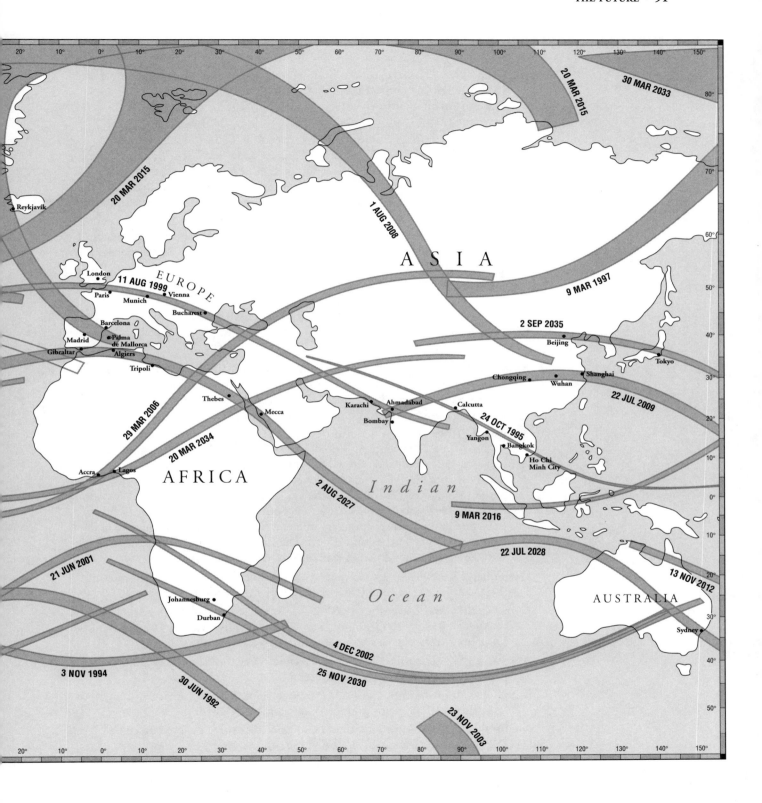

Notable Future Eclipses

June 30, 1992—The next total eclipse following July 11, 1991, barely touches land just east of Montevideo, Uruguay, before it passes out across the ocean. A south Atlantic cruise in June is a very chilly affair.

May 10, 1994—An *annular eclipse* path 140 miles wide passes over North America, from Baja California to Nova Scotia. With up to 6 minutes of annularity visible in cities such as El Paso, Oklahoma City, Detroit, Cleveland, and Buffalo, there will be grave danger of many people, especially schoolchildren, damaging their eyesight by looking directly at this "ring of Sun." (The path grazes Kansas City on the north and St. Louis on the south.)

November 3, 1994—Later that year a mid-autumn total eclipse passes largely over unpopulated areas of South America (it misses Machu Picchu by about 175 miles). Asunción, Paraguay, lies near the southern edge of the path.

October 24, 1995—This narrow path of totality (less than 50 miles wide) passes just a few miles south of Calcutta and out across the mouths of the Ganges River. It also passes less than 100 miles north of Yangon (Rangoon), Bangkok, Phnom Penh, and Ho Chi Minh City (Saigon).

March 9, 1997—Although the temperature for this Siberian eclipse will be very cold, the chances for clear skies are very good.

February 26, 1998—In late morning the Moon's shadow crosses the northern 40 miles of Isla Isabela, the largest of the Galapagos Islands, including the Wolf Volcano (5600 feet). Later, the umbra covers Maracaibo, Venezuela, as well as Antigua, Montserrat, and part of Guadeloupe in the West Indies.

August 11, 1999—"The Great European Eclipse of 1999" is the final total solar eclipse of the 20th century. The shadow crosses the English Channel, darkening Plymouth, England, and Cherbourg, France, but passing south of Stonehenge by some 60 miles. The list of cities in the path—Rouen, Reims, Luxembourg, Stuttgart, Munich, Salzburg—reads like a travelogue of Western Europe. (Paris and Vienna are each just outside the path of totality.) From there the umbra crosses Hungary and Romania (including Bucharest) and continues across Turkey and the Middle East, darkening both Karachi, Pakistan, and Ahmadabad, India, late on this Wednesday afternoon.

December 25, 2000—A *partial solar eclipse*, visible from the entire 48 states of the U.S., graces the final Christmas Day of the second millennium.

June 21, 2001 and **December 4, 2002**—A forty-mile stretch of the Atlantic coast of Angola, just north of Lobito, experiences totality twice in 18 months. The December eclipse passes 50 miles south of Victoria Falls.

April 8, 2005—The total phase of this *annular-total eclipse* never touches land, passing entirely over the Pacific Ocean.

March 29, 2006—Although this 110-mile wide path crosses Africa and much of Asia, there are no major population centers in its path.

August 1, 2008—Summer in Siberia may appeal to some eclipse chasers. This one also crosses the Great Wall of China near sunset.

July 22, 2009—The next eclipse in the July 11, 1991, *saros* series begins in the Arabian Sea just off the coast of India between Bombay and Ahmadabad. The path traverses central India (including Indore, Benares, and Patna) and then the Himalayas through eastern Nepal and Bhutan, passing 75 miles south of Mt. Everest. The shadow engulfs the Chinese cities of Chongqing, Wuhan, and Shanghai before passing out into the East China Sea and halfway across the Pacific.

July 11, 2010—This South Pacific path of totality passes within 15 miles of Tahiti (great cruise ship opportunities) and completely engulfs Easter Island for 4 minutes and 45 seconds of totality.

November 13, 2012—The path begins near Darwin in north central Australia and moves across the Great Barrier Reef near Cairns.

August 21, 2017—Finally, another total eclipse in the U.S. This one sweeps a 70-mile wide path from Salem, Oregon, to Charleston, South Carolina, crossing coast-to-coast from mid-morning to early afternoon on this summer Monday. Points in the path include Mt. Jefferson (Cascades), Grand Teton peak (Rockies), Casper, Wyoming, and Lincoln, Nebraska. In Missouri, Kansas City lies near the southern edge of the path, St. Louis near the northern edge. The eclipse reaches its maximum in western Kentucky, with the Sun at an altitude of 64° and a duration of totality of 2 minutes and 40 seconds. In Tennessee, Nashville is near the southern edge of the path, Knoxville near the northern edge. The shadow then passes over Greenville, Columbia, and Charleston, South Carolina, before racing out across the Atlantic.

April 8, 2024—Another North American total eclipse only seven years later! This 120-mile wide path crosses Mazatlán, Dallas, Cleveland, and Buffalo with more than four minutes of totality.

August 12, 2026—An unusual eclipse track that actually moves westward near the North Pole before swinging southeast over Greenland, Iceland, and Spain. Reykjavík, Madrid, and Barcelona are near the edge of the path, and Palma de Mallorca is near the centerline just before sunset.

August 2, 2027—The second eclipse in the same *saros* following July 11, 1991, passes over the Rock of Gibraltar, Thebes in Egypt (near where it reaches maximum duration of 6 minutes 22 seconds), and Mecca in Saudi Arabia.

July 22, 2028—This, another long eclipse, passes from one end of Australia to the other (including Sydney) on this afternoon. Near sunset, the path crosses Dunedin on South Island, New Zealand.

March 20, 2034—This eclipse crosses Africa and the Middle East, including southern Kuwait in the path of totality.

September 2, 2035—A grand oriental eclipse, the path crosses Beijing and just grazes Tokyo on the southern edge of the path.

August 12, 2045—The third eclipse in the same *saros* following July 11, 1991, sweeps across the United States from Northern California to Florida, as well as Haiti and the northeast coast of South America, including the mouth of the Amazon River. After three *saros* periods (54 years and about one month), the point of maximum eclipse (path not shown on map) has shifted about 4° north and has returned to a position about 27° east of the 1991 event.

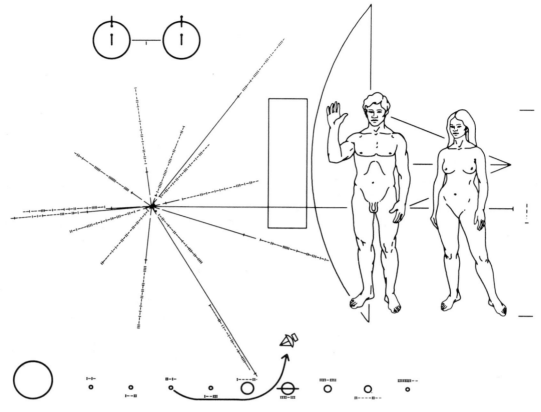

Plaque attached to Pioneer 10 spacecraft (1972)

From the Stone Age to the Space Age

Eclipses are one of the few types of events that we can predict with certainty centuries into the future. The surety that the Moon's shadow will pass over the Earth at a certain time and place somehow confers a reassuring sense of continuity to our fragile existence as a race of human beings. Ancient cultures embodied this cosmic order in stone monuments, clay tablets, and hieroglyphic books that have survived for thousands of years. A recent experiment reflects our ongoing efforts to preserve a record of our existence on Earth.

In 1987, the Pioneer 10 space probe became the first human-made object to leave the solar system. It passed beyond Pluto's orbit and entered the realm of deep space, traveling at a speed of seven miles per second. Before Pioneer 10 was launched in 1972 to study Jupiter, astronomers Carl Sagan and Frank Drake persuaded NASA to attach to the spacecraft a plaque designed to identify its origin to any intelligent extraterrestrial beings who might intercept it. The figures and diagrams etched on the gold plaque convey nonverbal information about the solar system, its location in the Galaxy, and the inhabitants of the third planet, Earth.

William Blake's "The Serpent Temple," *from* Jerusalem (1820)

The two human figures represent typical earthlings. Their size is shown in comparison to an outline of the Pioneer 10 spacecraft in the background. The two circles at the top represent atoms of hydrogen, the most abundant element in the universe. The starburst pattern shows the direction and distances to specific stars. At the bottom of the plaque are the Sun and the planets. Pioneer 10 is shown on its path leaving Earth, swinging by Jupiter, and passing forever beyond our solar system.

We have come a long way from stone monuments to interplanetary space probes. Yet our fascination with the Sun, the Moon, the stars, and the planets remains. Early attempts to understand the heavens were based on astrology: the supposed prediction of earthly events based on the positions of the planets. It was against this background that modern astronomy has developed. As we increase our knowledge of the universe, new patterns of reality will emerge. That is the destiny of science. Our expansion into the solar system is already bringing us exciting new views of other planets and other moons. But none of these planets or moons can produce an eclipse like we see here on Earth; this perfect matching of our lunar and solar disks is unique. It's this remarkable coincidence in time and space that gives us the experience we call a total eclipse of the Sun.

Glossary

altitude · the angle (in degrees) above the level horizon where an object in the sky appears. (The object's *azimuth* is also needed to pinpoint its position.)

annular eclipse · a solar eclipse that occurs when the apparent size of the Moon is not great enough to completely cover the Sun. A thin ring of sunlight can be seen around the black disk of the Moon.

annular-total eclipse · a solar eclipse that has both annular and total phases. (Also called a *central eclipse*.)

anomalistic month · the time it takes for the Moon to travel from apogee to perigee and back again (about 27.6 days).

aphelion · the point in the Earth's orbit that is farthest from the Sun. Currently the Earth reaches aphelion in early July.

apogee · the point in the Moon's orbit that is farthest from the Earth.

ascending node · the point in the orbit of the Moon where it passes from below the ecliptic plane to above. (See *node.*)

Aubrey holes · the 56 chalk-filled holes (named for John Aubrey) that mark the outer ring of Stonehenge. These holes may have served as "counters" to help in marking the cycles needed to predict eclipses.

azimuth · the compass direction (in degrees) where an object in the sky appears. (The object's *altitude* is also needed to pinpoint its position.)

Baily's beads · the effect seen just before and just after totality when only a few points of sunlight are visible at the edge of the lunar disk.

canon · in ancient times, an historical record of events. In modern astronomy, a canon is a listing of celestial events, such as eclipses, over a period of time.

central eclipse · in some references, a central eclipse refers to an eclipse that has both annular and total phases. (See *annular-total eclipse.*)

chromosphere · the lower atmosphere of the Sun that appears as a thin rosy ring around the edge of the solar disk during a total eclipse.

corona · the upper atmosphere of the Sun that appears as a halo around the Sun during a total eclipse.

contact · one of the instances when the apparent position of the edges of the Sun and the Moon cross one another during an eclipse. They are designated as first contact, second contact, third contact, and fourth contact.

descending node · the point in the orbit of the Moon where it passes from above the ecliptic plane to below. (See *node.*)

draconic month · the time it takes for the Moon to return to the same node (about 27.2 days).

eclipse · the alignment of celestial bodies so that one is obscured, either partially or totally, by the other.

eclipse season · the period of time when the Sun is near alignment with a lunar node, during which eclipses may take place. For solar eclipses, this 37½-day time window recurs every 173 days.

eclipse year · the length of time it takes for a lunar node to return to its original alignment with respect to the Sun (about 346.6 days).

ecliptic · the plane of the Earth's orbit around the Sun. As seen from the Earth, the Sun appears to move across the ecliptic during one year.

equinox · either of the two days when the periods of daylight and darkness are of equal length. The vernal equinox is usually March 21; the autumnal equinox is usually September 23.

first contact · the beginning of a solar eclipse marked by the edge of the Moon first passing across the disk of the Sun.

fourth contact · the end of a solar eclipse marked by the disk of the Moon completely passing away from the disk of the Sun.

G.M.T. · Greenwich Mean Time. A world time standard based on the meridian (0° longitude) that lies on a point in Greenwich, England.

heel stone · the large upright boulder (or menhir) at Stonehenge that is aligned with the summer solstice sunrise.

latitude · distance on the Earth (measured in degrees) north or south of the equator.

longitude · distance on the Earth (measured in degrees) east or west from a reference line, usually the line running between the poles passing through Greenwich, England.

lunar eclipse · the passage of the Moon into the shadow of the Earth, always occurring at a full Moon.

negative shadow · the extension of the umbra of an annular eclipse that delineates the path from which observers may see the ring of Sun of the annular eclipse.

node · the two points where a tilted orbit intersects a geometric plane. The Moon's orbit intersects the ecliptic plane at the ascending node and the descending node.

partial eclipse · an eclipse during which only the partial shadow touches the Earth (for a solar eclipse) or the Moon (for a lunar eclipse).

penumbra · the part of a shadow (as of the Moon) within which the source of light (the Sun) is only partially blocked out.

perigee · the point in the orbit of the Moon that is closest to the Earth.

perihelion · the point in the orbit of the Earth that is closest to the Sun. Currently the Earth reaches perihelion in early January.

prominence · a large-scale gaseous formation above the surface of the Sun.

regression · the movement of points in an orbit in the direction opposite from the motion of the orbiting body. For example, the Moon travels from west to east, but its nodes are regressing from east to west.

saros · the eclipse cycle with a period of 223 synodic months, or 6,585.32 days (18 years and about 11 days).

second contact · the beginning of the total phase of a solar eclipse marked by the leading edge of the Moon first completely obscuring the Sun.

shadow bands · faint ripples of light sometimes seen on flat, light-colored surfaces just before and just after totality.

solar eclipse · the passage of the new Moon directly between the Sun and the Earth when the Moon's shadow is cast upon the Earth. The Sun appears in the sky either partially or totally covered by the Moon.

solstice · the day when the noontime Sun is either highest in the sky (summer solstice is June 22) or lowest in the sky (winter solstice on December 22).

spectroscope · a scientific instrument that breaks light into its component wavelengths for measurement.

sunspot · a magnetic disturbance on the Sun that appears as a dark blotch on its surface.

synodic month · the time from one full Moon to the next (about 29.5 days).

third contact · the end of the total phase of a solar eclipse marked by the trailing edge of the Moon first revealing the Sun.

total eclipse · an eclipse during which the Moon's umbra touches the Earth (for a solar eclipse) or the Earth's umbra completely engulfs the Moon (for a lunar eclipse).

totality · the period during a solar eclipse when the Sun is completely blocked by the Moon. (Totality for a lunar eclipse is the period when the Moon is in the complete shadow of the Earth.)

umbra · a complete shadow (as of the Moon) within which the source of light (the Sun) is totally hidden from view.

zodiac · the division of the ecliptic plane into twelve equal parts; each of these parts or "signs" is identified by a name and symbol (for example, Sagittarius ♐).

References

General Reading

Menzel, Donald H., and Pasachoff, Jay M., *A Field Guide to the Stars and Planets,* 2nd ed., Houghton Mifflin, Boston (1990). This guide includes all kinds of information (plus maps) to help you find objects in the sky, including eclipses.

Mitchell, Samuel A., *Eclipses of the Sun,* Columbia Univ. Press, New York (Fifth edition, 1951). This basic text on solar eclipses blends the author's personal experiences with the history and science of eclipses.

Sweetsir, Richard, and Reynolds, Michael, *Observe: Eclipses,* Astronomical League, Washington, DC (1979). This handy field guide offers many practical tips on observing an eclipse.

Todd, Mabel L., *Total Eclipses of the Sun,* Little, Brown, & Co., Boston (1900). This popular treatment of the subject was produced in anticipation of the May 28, 1900, eclipse in the United States.

Zirker, J.B., *Total Eclipses of the Sun,* Van Nostrand Reinhold, New York (1984). This text focuses on the science of eclipses and the physics that can be learned from them.

National Geographic has included a number of articles on solar eclipses. See the issues of August 1970, November 1963, March 1949, September 1947, September 1937, February 1937, and November 1932.

Two monthly magazines, *Sky and Telescope* and *Astronomy,* provide much useful information. Hardly a month passes without some treatment of eclipses and related subjects.

Eclipse Data

The Astronomical Almanac, U. S. Government Printing Office, Washington, D. C. 20402. This annual volume contains (among much other information) the details of all eclipses for the year.

Espenak, Fred, *Fifty Year Canon of Solar Eclipses: 1986–2035,* Sky Publishing, Cambridge, MA (1987). This reference is unsurpassed for accurate and detailed information concerning upcoming solar eclipses.

Kudlek, Manfred, and Mickler, Erich H., *Solar and Lunar Eclipses of the Ancient Near East,* Verlag Butzon & Bercker Kevelaer, Hamburg (1971). This book gives eclipse data and maps from 3000 B.C. to the year 0 for important historical places in the ancient Near East.

Meeus, J., Grosjean, C. C., and Vanderleen, W., *Canon of Solar Eclipses*, Pergamon Press, Oxford (1966). This canon contains the worldwide data and maps of all solar eclipses between 1898 and 2510.

Oppolzer, Theodor R. von, *Canon of Eclipses*, Dover Publications, New York (1962). This is a reprint of Oppolzer's classic of 1887; it presents data and maps for eclipses from 1207 B.C. to 2161 A.D.

Solar Eclipse Photography

Lowenthal, James, *The Hidden Sun: Solar Eclipses and Astrophotography*, Avon, New York (1984). Much of this book deals with practical details about eclipse photography, and includes the author's personal anecdotes.

Astrophotography Basics, Kodak Customer Service Pamphlet Number AC-48 (1978). This concise and informative guide has much good information on eclipse photography.

Ancient Astronomy

Aveni, Anthony F., *Skywatchers of Ancient Mexico*, University of Texas Press, Austin (1980). This text includes a lengthy discussion of eclipse tables found in the Dresden Codex of the ancient Mayans.

Hawkins, Gerald S., *Stonehenge Decoded*, Doubleday, Garden City (1965). This is the original account of the discovery that Stonehenge may have been used to predict eclipses.

Hoyle, Fred, *On Stonehenge*, W. H. Freeman & Co., San Francisco (1977). This book delves deeper into the astronomical aspects of this ancient monument.

Other

Lineman, Rose, *Eclipses: Astrological Guideposts*, American Federation of Astrologers, Inc., Tempe, Arizona (1984). This book includes eclipse interpretations for those who are astrologically inclined.

Schatz, Dennis, *Astronomy Activity Book*, Simon and Schuster, New York (1991). This book provides numerous astronomy activities geared for school-age children.

Sonneborn, Ruth A., *Someone is Eating the Sun*, illustrated by Eric Gurney, Random House, New York (1974). This delightful 32-page children's book tells a story of animal characters reacting to a total solar eclipse.

Index

About the Author

Bryan Brewer wrote the first edition of *Eclipse* in anticipation of the February 26, 1979, total solar eclipse that passed near his home in Seattle, Washington. As he says: "It was a classic case of 'I couldn't find a good book on the subject, so I wrote one'." Now twelve years later, he looks forward to again enjoying the experience of totality, this time on Hawaii with his wife and two sons. No stranger to cosmic coincidences, Brewer cites two inspirations for his celestial inclinations: his older son, Devin, celebrated his second birthday on the February 26th eclipse; and his younger son, Matthew, celebrated his fifth birthday on April 11, 1986, the day Halley's Comet made its closest approach to Earth.

Brewer's broad interests in science, mathematics, geography, history, and psychology give him the background to write engagingly about all aspects of eclipses. He is the co-author of *The Compact Disc Book* and performs consulting work in the area of CD-ROM technology. He is also the producer of the audio compact disc *The Fine Art of Relaxation* by Joel Levey. The log house near Mt. Rainier (mentioned in this section of the first edition) has been completed, and after living there for a six-year period, the Brewers now enjoy the home as a vacation retreat.